Information Systems Engineering and Management

Volume 34

The book series "Information Systems Engineering and Management" (ISEM) publishes innovative and original works in the various areas of planning, development, implementation, and management of information systems and technologies by enterprises, citizens, and society for the improvement of the socio-economic environment.

The series is multidisciplinary, focusing on technological, organizational, and social domains of information systems engineering and management. Manuscripts published in this book series focus on relevant problems and research in the planning, analysis, design, implementation, exploration, and management of all types of information systems and technologies. The series contains monographs, lecture notes, edited volumes, pedagogical and technical books as well as proceedings volumes.

Some topics/keywords to be considered in the ISEM book series are, but not limited to: Information Systems Planning; Information Systems Development; Exploration of Information Systems; Management of Information Systems; Blockchain Technology; Cloud Computing; Artificial Intelligence (AI) and Machine Learning; Big Data Analytics; Multimedia Systems; Computer Networks, Mobility and Pervasive Systems; IT Security, Ethics and Privacy; Cybersecurity; Digital Platforms and Services; Requirements Engineering; Software Engineering; Process and Knowledge Engineering; Security and Privacy Engineering, Autonomous Robotics; Human-Computer Interaction; Marketing and Information; Tourism and Information; Finance and Value; Decisions and Risk; Innovation and Projects; Strategy and People.

Indexed by Google Scholar. All books published in the series are submitted for consideration in the Web of Science.

For book or proceedings proposals please contact Alvaro Rocha (amrrocha@gmail.com).

Narasimha Rao Vajjhala · Edlira Martiri ·
Fisnik Dalipi · Bian Yang
Editors

Artificial Intelligence in Healthcare Information Systems—Security and Privacy Challenges

Springer

Editors
Narasimha Rao Vajjhala
American University in Bulgaria
Blagoevgrad, Bulgaria

Fisnik Dalipi
Department of Informatics
Linnaeus University
Kalmar, Sweden

Edlira Martiri
Department of Statistics and Applied
Informatics, Faculty of Economy
University of Tirana
Tirana, Albania

Bian Yang
Norwegian University of Science
and Technology
Gjøvik, Norway

ISSN 3004-958X ISSN 3004-9598 (electronic)
Information Systems Engineering and Management
ISBN 978-3-031-84403-4 ISBN 978-3-031-84404-1 (eBook)
https://doi.org/10.1007/978-3-031-84404-1

This Springer imprint is published by the registered company Springer Nature Switzerland AG
The registered company address is: Gewerbestrasse 11, 6330 Cham, Switzerland

Foreword

Artificial intelligence (AI) stands as one of the most transformative forces in modern healthcare, reshaping everything from diagnostics to patient management with unprecedented speed and precision. However, the very qualities that make AI so promising also introduce new complexities, especially around the security and privacy of sensitive healthcare data. *Artificial Intelligence in Healthcare Information Systems—Security and Privacy Challenges* brings together a comprehensive collection of insights from researchers, technologists, and healthcare practitioners who are pioneering solutions to these pressing challenges. As the book's contributors explore AI applications across multiple healthcare domains, they underscore the importance of a secure, ethical, and collaborative approach to AI adoption.

This volume opens with foundational discussions on AI's potential within healthcare, particularly in diagnostics, disease prediction, and patient care. However, the authors do not shy away from the critical issues. They provide an in-depth look at the security challenges unique to healthcare information systems, where data breaches and cyberattacks can compromise not just privacy but also patient safety. Through chapters on encryption techniques, regulatory frameworks, and emerging data protection methodologies, this book equips readers with actionable strategies to protect patient data in an increasingly digital healthcare ecosystem. One of the book's standout aspects is its commitment to exploring AI's impact on specialized fields, such as fetal medicine and reproductive health, areas that demand high accuracy and confidentiality. Through a mix of technical depth and practical insights, the authors illustrate how AI can optimize clinical workflows, reduce errors, and support clinicians in making informed decisions. These case studies from diverse settings, including countries like India and Kazakhstan, reflect a truly global approach to AI in healthcare, recognizing the varied challenges and resources different regions bring to this transformative field.

Equally significant is the book's attention to the ethical considerations and biases that may arise in AI-driven healthcare applications. As algorithms become more deeply embedded in clinical decision-making, questions of fairness, transparency, and accountability take on greater urgency. The authors advocate for unbiased AI models and suggest ways to mitigate data biases, thereby ensuring that AI technologies serve all patients equitably. This emphasis on ethics is crucial in building trust between healthcare providers, patients, and AI systems, and it aligns with a broader vision of AI as a tool for inclusive, high-quality healthcare. Data protection is another focal point, with the book offering a thorough review of encryption methods and advanced security protocols tailored for healthcare. From homomorphic encryption to blockchain applications, the authors present a variety of solutions that can help institutions secure their data without compromising accessibility or functionality. These approaches are discussed alongside existing regulatory frameworks such as HIPAA and GDPR, illustrating how legal and technological measures can work together to create a safe environment for AI-enabled healthcare.

In addition to security, this book explores the practical challenges of implementing AI in resource-limited settings, where infrastructure and training may be lacking. By examining AI's potential within the context of diverse healthcare environments, the book broadens the scope of what AI can achieve, highlighting real-world applications in both advanced and emerging healthcare systems. The insights provided here serve as a guide for stakeholders to adopt AI solutions that are both sustainable and sensitive to the unique demands of their local healthcare landscapes. The perspectives of healthcare providers are particularly valuable in this volume, as they offer firsthand insights into the practicalities of AI adoption in clinical settings. By focusing on training, usability, and provider acceptance, the book emphasizes the human side of AI in healthcare, recognizing that successful implementation requires more than just technology—it requires buy-in and adaptability from those who use these tools daily. Such discussions ensure that the book's approach remains holistic, addressing the technological, ethical, and human dimensions of AI in healthcare.

Big data and telehealth are also covered extensively, presenting AI as a means to expand access to healthcare and improve patient outcomes. These technologies, while promising, introduce their own set of challenges, particularly around data interoperability and quality. The book's discussion on these topics is forward-looking, considering how AI can not only improve healthcare delivery today but also lay the groundwork for a more integrated and data-driven healthcare future. Ultimately, *Artificial Intelligence in Healthcare Information Systems—Security and Privacy Challenges* is a timely and essential resource for anyone involved in the intersection of healthcare and technology. It provides healthcare professionals, IT specialists, policymakers, and researchers with the insights and tools they need to navigate the complexities of AI adoption responsibly. By offering a balanced view that considers both opportunities and risks, the book equips its readers to make informed, ethical decisions in the pursuit of enhanced patient care.

As we stand on the cusp of a new era in healthcare, one driven by AI and data innovation, the lessons and strategies contained in this book are more relevant than ever. It is a call to action for the healthcare community to embrace AI thoughtfully, ensuring that technological advancements go hand in hand with robust security and ethical standards. This book serves as a guide and an inspiration, reminding us of the immense possibilities AI holds for transforming healthcare—safely, securely, and for the benefit of all.

Assoc. Prof. Adeyemi Abel Ajibesin
Cape Peninsula University
of Technology
Cape Town, South Africa

Adeyemi Abel Ajibesin is a distinguished academic scholar with over two decades of extensive experience in research, teaching, and leadership, particularly in Electrical Engineering and Computer Science. Currently serving as a visiting professor in the Department of IT at Cape Peninsula University of Technology in South Africa, he previously held positions such as Chair/HoD of Software Engineering at the School of IT and Computing (SITC) at the American University of Nigeria, founding Interim Dean of the School of Engineering, Interim Dean of the School of IT and Computing, and Director of the African Centre for ICT Innovation and Training. He has also served in various capacities at Simon Fraser University in Canada, the University of Cape Town in South Africa, and the Pan African University Institute for Basic Sciences, Technology Innovation in Kenya, among other institutions. His areas of research are data science, broadband networks, frontier analysis, and AI. His prolific research portfolio is reflected in numerous articles published in reputable peer-reviewed conference proceedings, journals, and books, alongside patents for engineering products and copyrighted computing software, showcasing his innovative contributions to his field. Dr. Ajibesin's leadership and academic excellence have been recognized through numerous awards and grants, including esteemed accolades such as the Google Research (exploreCSR), AIMS, CSIR, and IEEE.

Preface

In recent years, artificial intelligence (AI) has rapidly transformed the healthcare landscape, introducing groundbreaking advancements in diagnostics, treatment planning, and patient management. However, this accelerated integration of AI into healthcare information systems has also surfaced critical security and privacy challenges that demand urgent attention. *Artificial Intelligence in Healthcare Information Systems—Security and Privacy Challenges* explores these complexities, providing a comprehensive analysis of how AI can be leveraged effectively while addressing the inherent risks associated with handling sensitive healthcare data. This book underscores the dual responsibility of healthcare professionals and technologists to enhance patient care with AI innovations while implementing robust data protection measures and ethical safeguards.

This book offers a perspective on AI's potential to revolutionize healthcare across diverse areas, from predictive diagnostics to advanced applications in fetal medicine, coupled with real-world case studies from countries like India and Kazakhstan. By examining both the technological advancements and the privacy frameworks necessary for secure AI integration, this book equips readers with a well-rounded understanding of the challenges and opportunities AI brings to healthcare. This book serves as a crucial resource for healthcare practitioners, IT specialists, and policymakers dedicated to navigating the future of AI in healthcare securely and responsibly. Through a focus on encryption, data protection, and ethical considerations, this book provides essential guidance for ensuring that AI-driven healthcare solutions respect and uphold patient privacy.

The opening chapter, titled "Artificial Intelligence in Healthcare: Navigating Security and Privacy Challenges," by Eriona Çela, Mathias Fonkam, Alexey Vedishchev, and Narasimha Rao Vajjhala, offers an in-depth exploration of AI's transformative potential within the healthcare sector. This chapter begins by analyzing how AI applications, such as machine learning, natural language processing, and computer vision, have enhanced diagnostics, personalized treatment, and improved operational efficiency. The authors emphasize AI's substantial impact on diagnostic accuracy, particularly in fields like radiology, where algorithms detect medical anomalies with precision. However, with these advancements come serious security and privacy concerns, as AI systems handle vast amounts of sensitive patient data. The chapter provides a critical overview of regulatory standards like HIPAA and GDPR, which govern data use in healthcare, and discusses essential data protection techniques, including encryption, anonymization, and privacy-preserving methods like federated learning and differential privacy. This chapter also tackles ethical considerations related to AI in healthcare, such as the importance of algorithmic transparency and the need for explainable AI to promote trust among both healthcare providers and patients. Through discussions on cutting-edge solutions like blockchain and quantum cryptography, the authors highlight potential future directions for secure AI applications in healthcare. Concluding with a call for adaptive regulatory frameworks, this chapter establishes a strong foundation for understanding the balance between leveraging AI for medical innovation and maintaining stringent safeguards for patient privacy and data security.

The second chapter, titled "AI in Diagnostic and Disease Prediction" by Hermehar P. S. Bedi, Renu Puri, Amandeep Kaur, and Deepak Puri, examines the transformative role of AI in early disease detection, with a particular focus on cancer. This chapter emphasizes the importance of timely diagnosis in reducing cancer mortality rates, which remains a leading cause of death globally. Highlighting advancements in AI and machine learning (ML), the authors explore how these technologies have revolutionized traditional diagnostic methods, enhancing accuracy and efficiency. Through an analysis of imaging techniques like MRI, PET, and CT scans, the chapter illustrates how AI-based models reduce reliance on manual interpretation and streamline the diagnostic process, ultimately improving patient outcomes. In addition, this chapter provides an in-depth review of key AI-ML techniques such as deep learning models and supervised ML algorithms, detailing their application in cancer and disease prediction. The authors discuss the strengths of algorithms like convolutional neural networks (CNNs), support vector machines (SVMs), and random forests (RFs), along with emerging hybrid models that offer increased precision. By presenting a comprehensive literature review and discussing new trends, the chapter emphasizes AI's potential to redefine healthcare diagnostics and disease prevention strategies, while also considering the ethical and technical challenges associated with widespread implementation. This chapter underscores the book's emphasis on AI's potential to enhance patient outcomes while addressing the accuracy and data security challenges involved in healthcare diagnostics.

The third chapter, titled "Comprehensive Analysis of Machine Learning Approaches for Identifying Polycystic Ovary Syndrome" by Mehak Abbas and Shaheen Naz, addresses the growing incidence of PCOS, a significant hormonal disorder affecting women's reproductive health. This chapter explores the critical role of machine learning (ML) in early detection and diagnosis of PCOS, presenting various ML models such as Convolutional Neural Networks (CNN), Support Vector Machines (SVM), Random Forest (RF), and Decision Trees (DT). The authors review commonly used datasets in PCOS studies, comparing the effectiveness of these algorithms and highlighting the advantages and limitations of each approach. Through extensive analysis, this chapter underscores the value of ML in improving diagnostic accuracy and provides insights into the challenges and potential future advancements in ML-based PCOS detection. This chapter exemplifies the application of machine learning in specific healthcare scenarios, contributing to the book's aim to explore AI's precision in identifying health conditions while balancing patient data sensitivity.

The fourth chapter, "Artificial Intelligence: A New Paradigm in Fetal Medicine Research and Clinical Practice," by Sharan Kumar R., Raghesh Krishnan K., Neha Garg, and Reema Bhatt, examines the transformative applications of AI in fetal medicine. This chapter emphasizes the integration of AI techniques, including Deep Learning (DL) and Convolutional Neural Networks (CNN), in prenatal ultrasound to automate standard measurements, enhance imaging accuracy, and facilitate early diagnosis of foetal abnormalities. The authors discuss the impact of AI on improving biometric precision and reducing sonographer workload, offering case studies that demonstrate AI's potential in advancing foetal health outcomes. Ethical considerations and data privacy challenges are also explored, underscoring the balance between technological innovation and responsible clinical application. This chapter aligns with the book's theme of transforming specialized medical fields with AI while maintaining secure and ethical data handling practices.

The fifth chapter, titled "Trends in AI for Disease and Diagnostic Prediction: A Healthcare Perspective," by Saurav Kumar Mishra, Tabsum Chhetri, Anagha Balakrishnan, Sneha Roy, Kusum Gurung, Jeba Praba J., and John J. Georrge, provides an in-depth review of the role of AI in transforming healthcare diagnostics. This chapter highlights AI's capability to analyze large datasets for early disease prediction, enabling proactive interventions in areas like oncology, cardiovascular health, and neurology. The authors also examine various AI technologies, including Natural Language Processing (NLP), Deep Learning (DL), and computer vision, showcasing their applications across multiple medical disciplines. Ethical concerns, patient privacy, and the need for interpretability in AI-driven insights are discussed to address the broader challenges and future opportunities AI presents in healthcare. This chapter addresses the broader trends in AI for healthcare, reinforcing the book's focus on AI's role in disease prediction and the associated ethical considerations for patient privacy.

The sixth chapter, titled "Exploring the Impact of Perceived Entrepreneurial Mindset and AI Adoption on Customer Experiences: A Study of Just Ask Chatbot Users," by Shumayela Hasan and Richi Simon, examines the role of an entrepreneurial mindset in adopting AI technologies and enhancing customer experience within the healthcare sector. This mixed-methods study investigates the Just Ask Chatbot, an AI-powered tool designed to provide accessible information on adolescent healthcare and reproductive rights. Using a sample of 58 users, the authors employed Structural Equation Modeling (SEM) for quantitative analysis and content analysis for qualitative insights. Findings reveal a positive correlation between a proactive entrepreneurial mindset and customer experience, suggesting that innovative, entrepreneur-driven approaches can significantly enhance user satisfaction and engagement with AI technologies. Additionally, this chapter examines the ethical considerations of AI adoption in sensitive areas like healthcare, emphasizing user privacy, data security, and ethical standards. This chapter discusses the chatbot's success in promoting safe, reliable access to healthcare information, linking users to essential resources while respecting data privacy. Through an application of E. M. Rogers' Diffusion of Innovation Theory, the authors outline how an entrepreneurial mindset influences AI adoption, particularly in startup environments. The chapter provides valuable insights for startup founders and industry stakeholders, advocating for ethical, innovation-driven practices to achieve sustainable growth and competitive advantage. Through a case study on AI-driven healthcare tools, this chapter connects to the book's scope by examining the relationship between innovation and user experience, especially in ethically sensitive healthcare applications.

The seventh chapter, titled "Ethical Considerations and Bias in AI Applications in Healthcare" by Azamat Ali, provides a thorough examination of the ethical issues and potential biases inherent in implementing AI technologies within healthcare. This chapter discusses key ethical challenges, such as patient privacy, informed consent, transparency, and accountability in AI-driven medical decisions. Highlighting the risks of algorithmic bias arising from imbalanced data and flawed design, the chapter emphasizes the importance of fair and unbiased AI models to avoid disparities in healthcare delivery. Through a patient-centered approach, the author advocates for ethical safeguards to ensure equitable treatment and minimize bias. The chapter also explores various types of bias—data, algorithmic, and human biases—and their impact on healthcare outcomes, underscoring real-world examples of gender, racial, and socioeconomic biases in AI algorithms. Ali proposes a multi-pronged strategy to address these issues, emphasizing the need for diverse datasets, transparent AI models, and ongoing ethical oversight. The chapter concludes by outlining recommendations to promote fairness and equal treatment in healthcare AI, urging stakeholders to adopt comprehensive strategies to uphold ethical standards while leveraging AI's potential to enhance patient care. This chapter directly addresses the book's core theme of ethical challenges, focusing on bias mitigation to ensure fair and unbiased AI applications within healthcare.

The eighth chapter, titled "A Comprehensive Review of Encryption and Protection Techniques for Healthcare Data" by Venkata Nedunoori, provides an in-depth analysis of modern encryption and data protection strategies essential for safeguarding sensitive healthcare information. This chapter explores foundational encryption principles, such as symmetric and asymmetric encryption, detailing their applications in protecting data at rest and in transit within healthcare systems. Key algorithms like Advanced Encryption Standard (AES) and Rivest-Shamir-Adleman (RSA) are examined for their effectiveness in various healthcare scenarios. The author also reviews advanced techniques, including Field-Level Encryption and End-to-End Encryption, which provide granular data security control, highlighting the balance between data utility and privacy. The chapter further examines emerging encryption technologies, such as homomorphic encryption, blockchain, and quantum cryptography, assessing their potential to address the unique security challenges posed by evolving healthcare data needs. With a focus on regulatory compliance, the author discusses the implications of standards like HIPAA and GDPR, emphasizing their role in shaping encryption practices. The chapter concludes with insights into future trends, including adaptive encryption systems and the integration of AI for automated data protection, presenting a forward-looking perspective on healthcare data security. Offering a detailed look at data protection methods, this chapter complements the book's emphasis on securing healthcare data in AI-driven systems to uphold patient privacy.

The ninth chapter, titled "Artificial Intelligence (AI) Adoption and Implementation Challenges in Indian Healthcare Sector" by Madhavi Kilaru and Rajasekhara Mouly Potluri, explores the transformative potential of AI in addressing the pressing healthcare needs of India's vast population. The chapter examines the ways AI can streamline administrative tasks, enhance diagnostic accuracy, optimize resource allocation, and improve patient management. In a nation with a high demand for healthcare services and limited healthcare professionals, AI offers innovative solutions to improve access and clinical outcomes. However, the chapter highlights the significant challenges to implementing AI, including infrastructural limitations, data privacy concerns, and the need to upskill the healthcare workforce. This chapter provides an in-depth look at various applications of AI in India's healthcare sector, such as virtual assistants, diagnostic assistance, drug discovery, telemedicine, predictive analytics, and remote patient monitoring. Each of these areas presents unique challenges, especially in terms of regulatory compliance, data security, and operational infrastructure. The authors emphasize the importance of building a supportive ecosystem with adequate infrastructure, regulatory frameworks, and workforce training to fully harness AI's potential in healthcare. The chapter concludes with recommendations for overcoming these obstacles, highlighting the need for a coordinated effort among government, healthcare providers, and technology developers to ensure safe, efficient, and ethical AI integration in the healthcare system. By exploring AI adoption within the Indian healthcare context, this chapter aligns with the book's examination of global challenges and strategies for integrating AI while safeguarding sensitive patient information.

The tenth chapter, titled "Problems and Prospects of AI in the Global Healthcare Sector: A Case of Kazakhstan" by Rajasekhara Mouly Potluri, Yerzhan Mukashev, and Inzhu Baktygulova, investigates the challenges and opportunities AI presents in Kazakhstan's healthcare landscape. This chapter leverages surveys and a combination of qualitative and quantitative methods to analyze the healthcare system, focusing on key issues such as limited infrastructure, disparities in rural healthcare access, and workforce shortages. The authors examine how AI technologies can address these challenges by enhancing diagnostic accuracy, optimizing resource distribution, and expanding telemedicine, particularly in underserved regions. The chapter also contextualizes Kazakhstan's healthcare issues within a global perspective, discussing how AI-driven solutions can be integrated despite data limitations and system complexities. By comparing Kazakhstan's healthcare dynamics with international models and exploring socioeconomic influences, the chapter provides valuable insights for policymakers and healthcare professionals seeking to harness AI's transformative potential. The authors conclude with recommendations for creating a supportive ecosystem that includes policy reforms, infrastructural improvements, and cross-sector collaboration to drive AI adoption in healthcare. This chapter offers a comparative international perspective, illustrating the challenges and opportunities of AI in diverse healthcare systems, aligning with the book's scope of assessing AI's global implications in healthcare.

The eleventh chapter, titled "Transforming Healthcare: The Role of Health Informatics and Provider Perspectives" by Philip Eappen, Ian Stedman, and Virginia Gunn, explores healthcare providers' perspectives on health informatics and the transformative potential it holds for patient care. This chapter outlines the benefits of health informatics, such as enhanced patient care, improved communication, and streamlined workflows, while also addressing persistent challenges, including data security concerns and the need for provider acceptance. Through a comprehensive literature review, the authors identify three critical areas—training, infrastructure, and institutional policies—that are essential for effective health informatics adoption. Emphasizing the importance of tailored training programs, robust technological infrastructure, and clear institutional guidelines, the chapter advocates for a holistic approach to integrating health informatics into clinical practice. Additionally, the chapter discusses the impact of big data, telehealth, and health information exchanges on healthcare delivery, analyzing how these innovations can improve efficiency and accessibility. By examining the integration of technologies like electronic health records (EHRs), telemedicine, and wearable devices, the authors highlight the potential for these tools to empower healthcare providers and patients. The chapter concludes by calling for ongoing collaboration among healthcare professionals, policymakers, and technology developers to create an environment that supports secure, efficient, and patient-centered healthcare delivery through health informatics. Focusing on health informatics and healthcare providers' perspectives, this chapter complements the book's theme by emphasizing the need for secure, user-friendly AI applications to improve healthcare delivery and patient data protection.

Artificial Intelligence in Healthcare Information Systems—Security and Privacy Challenges presents a diverse and comprehensive exploration of how AI is reshaping the healthcare sector, with a particular focus on the critical security and privacy aspects that must be addressed to safeguard sensitive patient data. Across its chapters, this book provides insights into the transformative potential of AI technologies in improving diagnostic accuracy, predictive analytics, and patient management, while also navigating the complexities of ethical considerations, regulatory compliance, and data protection. Each chapter offers unique perspectives from both technological and clinical viewpoints, creating a balanced discourse that reflects the challenges and opportunities of AI implementation in healthcare. The book serves as a valuable resource for healthcare professionals, IT specialists, researchers, and policymakers who are committed to driving innovation in healthcare responsibly and ethically.

As we move toward a future where AI will continue to play an even more integral role in healthcare, this book underscores the importance of enabling a collaborative ecosystem among medical practitioners, technologists, and regulators. Such an environment is essential to addressing the complex challenges of AI adoption while realizing its full potential to improve healthcare outcomes worldwide. We hope that readers will find the insights in this book both inspiring and practical, equipping them with the knowledge needed to navigate the rapidly evolving landscape of AI in healthcare and to contribute meaningfully to the development of secure, efficient, and ethically sound healthcare systems.

Blagoevgrad, Bulgaria — Narasimha Rao Vajjhala
Tirana, Albania — Edlira Martiri
Kalmar, Sweden — Fisnik Dalipi
Gjøvik, Norway — Bian Yang

Contents

Editors and Contributors

About the Editors

Narasimha Rao Vajjhala is an academic and researcher currently serving as Professor and Chair of the Computer Science Department at the American University in Bulgaria (AUBG). He is also the Principal Investigator for the Democracy Project (DemPro) at the Center for Information, Democracy, and Citizenship (CIDC) at AUBG. Dr. Vajjhala previously held leadership roles as Dean of the Faculty of Engineering and Architecture at the University of New York Tirana (UNYT) in Albania and Chair of the Computer Science and Software Engineering programs at the American University of Nigeria (AUN). With over 23 years of experience in academia, he has taught programming, database management, and related courses across Europe and Africa at both undergraduate and graduate levels. An active contributor to the global research community, Dr. Vajjhala is a Senior Member of ACM and IEEE, Editor-in-Chief of the International Journal of Risk and Contingency Management (IJRCM), and a member of the Project Management Institute (PMI). Beyond academia, he has worked as a technology consultant for European firms and has been actively involved in EU-funded projects. Dr. Vajjhala holds a Doctorate in Information Systems and Technology from the United States, a Master of Science in Computer Science and Applications from India, and a Master of Business Administration (MBA) specializing in Information Systems from Switzerland. His expertise spans computer science, project management, and information systems, making significant contributions to both research and industry.

Edlira Martiri is an Associate Professor at the University of Tirana, where she works as a researcher and lecturer. She has a background in Computer Science and Information Security and has been awarded two Ph.D. degrees from the University of Tirana, and NTNU, Norway. Both of her doctoral dissertations concentrated on data protection, with the first on images and the second on Biometric Systems and Machine Learning, highlighting her specialized knowledge in Information Security. Apart from her research work she has been involved in different roles in the industry where she has experience in threat analysis, attacker scenario modelling, risk management, and more.

Fisnik Dalipi holds a Ph.D. in Computer Science and is currently working as an associate professor at the Department of Informatics of Linnaeus University (LNU) in Sweden. Besides, he also holds the title of associate professor in information systems from the University of South-Eastern Norway, where he was previously working. His research embodies a robust interdisciplinary approach, leveraging computational approaches to address digitalization challenges in both industry and society. It bridges technology and societal needs, with a focus on areas like digital transformation and sustainable development. Dr. Dalipi's work emphasizes secure, efficient, and transformative data solutions to drive industry/society progress and innovation. His research applies AI to various domains, including security and privacy. Recently, he has been addressing blockchain and AI's role in improving privacy and security mechanisms in digital environments. He has been supervising students at different academic levels and has been involved in the development and curriculum preparation at all academic levels. He is actively serving as a program committee member and reviewer in many international conferences, workshops, and journals.

Bian Yang is a professor at the Norwegian University of Science and Technology (NTNU) in Gjøvik, specializing in biometrics, data privacy, and information security. He leads the eHealth and Welfare Security (eHWS) research group within NTNU's Department of Information Security and Communication Technology. His academic qualifications include degrees in electronic engineering and information security from the Harbin Institute of Technology, followed by research in biometric data protection, privacy-enhancing technologies, and identity management. Bian Yang has been involved in several EU and national research projects and has published widely on topics such as biometric template protection, cryptosystems, and healthcare information security. His work is recognized internationally, and he continues to contribute to projects aimed at enhancing security in healthcare and identity management systems.

Contributors

Mehak Abbas Department of Electrical and Electronics Communication Engineering, Sharda University, Greater Noida, India

Azamat Ali School of Liberal and Creative Arts, Lovely Professional University, Phagwara, Punjab, India

Inzhu Baktygulova KBTU Business School, Kazakh-British Technical University, Almaty, Kazakhstan

Anagha Balakrishnan Department of Bioinformatics, University of North Bengal, Siliguri, India

Hermehar P. S. Bedi Department of Human-Centered Design, Indraprastha Institute of Information & Technology, Delhi, India

Reema Bhatt Department of Fetal Medicine and Perinatology, Amrita Institute of Medical Sciences and Research Centre, Faridabad, Haryana, India

Eriona Çela University of New York Tirana, Tirana, Albania

Tabsum Chhetri Department of Bioinformatics, University of North Bengal, Siliguri, India

Philip Eappen Cape Breton University, Sydney, NS, Canada

Mathias Fonkam Penn State University, University Park, PA, USA

Neha Garg Department of Fetal Medicine and Perinatology, Amrita Institute of Medical Sciences and Research Centre, Faridabad, Haryana, India

John J. Georrge Department of Bioinformatics, University of North Bengal, Siliguri, India

Virginia Gunn Cape Breton University, Sydney, NS, Canada

Kusum Gurung Department of Bioinformatics, University of North Bengal, Siliguri, India

Shumayela Hasan The Bal Bhawan School, Bhopal, India

J. Jeba Praba Department of Computer Science and Applications, Christ College, Rajkot, India

Amandeep Kaur Department of Computer Science & Engineering, Chandigarh Engineering College—CGC Landran, Mohali, Punjab, India

Madhavi Kilaru VNR Vignana Jyothi Institute of Engineering and Technology, Hyderabad, India

Saurav Kumar Mishra Department of Bioinformatics, University of North Bengal, Siliguri, India

Yerzhan B. Mukashev KBTU Business School, Kazakh-British Technical University, Almaty, Kazakhstan

Shaheen Naz Department of Electrical and Electronics Communication Engineering, Sharda University, Greater Noida, India

Venkata Nedunoori Dentsu International, Frisco, TX, USA

Rajasekhara Mouly Potluri KBTU Business School, Kazakh-British Technical University, Almaty, Kazakhstan

Deepak Puri Doaba College of Pharmacy, Mohali, Punjab, India

Renu Puri Department of Computer Science & Engineering, Chandigarh Engineering College—CGC Landran, Mohali, Punjab, India

K. Raghesh Krishnan Department of Computer Science and Engineering, Amrita School of Computing, Amrita Vishwa Vidyapeetham, Coimbatore, India

Sneha Roy Department of Bioinformatics, University of North Bengal, Siliguri, India

R. Sharan Kumar Department of Computer Science and Engineering, Amrita School of Computing, Amrita Vishwa Vidyapeetham, Coimbatore, India

Richi Simon The Bhopal School of Social Sciences, Bhopal, India

Ian Stedman York University, Toronto, ON, Canada

Narasimha Rao Vajjhala American University in Bulgaria, Blagoevgrad, Bulgaria

Alexey Vedishchev American University in Nigeria, Yola, Nigeria

Artificial Intelligence in Healthcare: Navigating Security and Privacy Challenges

Eriona Çela, Mathias Fonkam, Alexey Vedishchev, and Narasimha Rao Vajjhala

Abstract This chapter examines the role of Artificial Intelligence (AI) in healthcare, examining its potential and the security and privacy challenges that accompany its integration. AI technologies, including machine learning, natural language processing, and computer vision, have significantly enhanced diagnostic accuracy, treatment personalization, and operational efficiency. However, the deployment of AI in healthcare also raises critical concerns around data security and patient privacy. As AI systems handle vast amounts of sensitive patient data, they are subject to rigorous regulatory standards such as the Health Insurance Portability and Accountability Act (HIPAA) and the General Data Protection Regulation (GDPR). The chapter discusses encryption, anonymization, and privacy-preserving techniques like federated learning and differential privacy, essential for maintaining data confidentiality. Additionally, it highlights ethical considerations, including algorithmic transparency and the need for explainable AI to promote trust among healthcare professionals and patients. The chapter concludes by exploring future directions in secure AI applications, such as blockchain and quantum cryptography, emphasizing the need for adaptive regulatory frameworks to support AI innovation while safeguarding patient rights. Through a balanced analysis, this chapter provides a comprehensive overview of the benefits and risks of AI in healthcare.

Keywords Artificial intelligence · Healthcare · Data security · Patient privacy · Machine learning · Natural language processing · Diagnostic accuracy · Federated learning · Blockchain · Regulatory compliance

E. Çela
University of New York Tirana, Tirana, Albania

M. Fonkam
Penn State University, University Park, PA, USA

A. Vedishchev
American University in Nigeria, Yola, Nigeria

N. R. Vajjhala (✉)
American University in Bulgaria, Blagoevgrad, Bulgaria
e-mail: nrao@aubg.edu

N. R. Vajjhala et al. (eds.), *Artificial Intelligence in Healthcare Information Systems—Security and Privacy Challenges*, Information Systems Engineering and Management 34, https://doi.org/10.1007/978-3-031-84404-1_1

1 Introduction

Artificial Intelligence (AI) has emerged as a transformative force in healthcare, significantly enhancing various aspects of medical practice, from diagnostics to patient management [1, 2]. The advancements in AI technologies over the past few years have led to their integration into clinical workflows, promising improved efficiency, accuracy, and patient outcomes. AI encompasses a range of technologies that simulate human intelligence, including machine learning, natural language processing, and computer vision. These technologies are increasingly being utilized in healthcare settings to optimize clinical decision-making, enhance diagnostic accuracy, and streamline operational processes. For instance, AI applications have been particularly effective in areas such as radiology, where algorithms can analyze medical images to detect anomalies with high precision [3, 4]. The integration of AI in diagnostic imaging has shown promise in improving workflow efficiency and reducing the workload on radiologists, thereby allowing them to focus on more complex cases [3, 5].

One of the most significant applications of AI in healthcare is in diagnostic imaging [6, 7]. AI-powered algorithms, particularly deep learning models, have shown great promise in analyzing medical images with a high degree of accuracy. Studies indicate that these algorithms can often match or exceed the diagnostic capabilities of human radiologists in identifying abnormalities such as tumors, fractures, and organ lesions [8]. For example, Convolutional Neural Networks (CNNs) have been particularly effective in interpreting complex imaging data, contributing to faster, more accurate diagnosis and treatment planning in fields such as radiology and oncology [9]. In recent years, AI has rapidly advanced, transforming healthcare with its ability to improve diagnostic accuracy, streamline processes, and enable personalized care. These advancements have emerged primarily through the development of machine learning algorithms, natural language processing, and predictive analytics, which are applied across a range of healthcare functions [10].

AI also plays a crucial role in predicting patient outcomes and guiding treatment decisions. Predictive analytics, which uses data from electronic health records (EHRs) and clinical databases, enables clinicians to anticipate complications, readmissions, and disease progression [11]. By analyzing patient history, genetics, and current health status, AI algorithms can suggest personalized treatment plans tailored to individual needs, improving patient outcomes and optimizing resource use in healthcare settings [12]. Natural language processing (NLP) represents another area of AI advancement, especially useful in interpreting unstructured medical data. NLP algorithms can extract valuable information from clinical notes, patient histories, and published medical research, making it accessible for clinical decision-making [13]. For instance, NLP has been successfully implemented in disease surveillance and monitoring systems to detect patterns and predict outbreaks, thereby supporting public health efforts [14]. AI has facilitated the development of robotic process automation (RPA) in healthcare administration [15]. RPA systems assist in automating repetitive tasks such as billing, appointment scheduling, and patient records management, thereby reducing administrative burdens on healthcare staff and allowing more time to focus on patient care [16]. This automation has proven beneficial in enhancing operational efficiency and reducing costs within healthcare systems.

AI has been instrumental in the management of chronic diseases and the personalization of treatment plans [17]. By analyzing vast amounts of patient data, AI systems can identify patterns and predict disease progression, enabling healthcare providers to tailor interventions to individual patient needs [18, 19]. This capability is particularly beneficial in oncology, where AI-driven decision support systems assist clinicians in selecting the most effective treatment options based on patient-specific data [20]. The potential of AI to enhance precision medicine is underscored by its ability to integrate diverse data sources, including genomic information and electronic health records, to inform clinical decisions [21].

Despite the promising advancements, the integration of AI into healthcare is not without challenges. One significant barrier is the resistance from healthcare professionals, often stemming from a lack of understanding of AI technologies and their implications for clinical practice. Studies have indicated that many healthcare workers are hesitant to adopt AI tools due to concerns about their reliability and the potential for job displacement [22, 23]. This resistance highlights the need for comprehensive training programs that equip healthcare professionals with the necessary skills to effectively utilize AI tools in their practice [21, 22]. Regulatory and ethical considerations pose additional challenges to the widespread adoption of AI in healthcare. The transparency of AI algorithms and the data on which they are trained are critical factors influencing their acceptance among both clinicians and patients [23]. Concerns regarding data privacy, algorithmic bias, and the ethical implications of AI decision-making must be addressed to promote trust in these technologies [23]. Regulatory bodies are beginning to establish guidelines for the development and implementation of AI applications in healthcare, but further efforts are needed to ensure that these technologies are both safe and effective [24, 25].

2 Benefits of AI in Healthcare

AI is rapidly transforming the healthcare landscape, offering innovative solutions across various domains, including diagnostics, treatment, and patient care [26, 27]. The integration of AI-driven tools into healthcare systems has the potential to enhance accuracy, efficiency, and personalization in patient management, ultimately improving health outcomes. In diagnostics, AI algorithms have demonstrated remarkable capabilities in analyzing medical images, such as X-rays, CT scans, and MRIs, often achieving accuracy levels comparable to or exceeding those of human experts [28]. For instance, AI-driven image analysis tools can identify anomalies in radiological images, facilitating early detection of conditions such as cancer and cardiovascular diseases [28, 29]. AI systems can assist in interpreting complex data sets, enabling healthcare professionals to make informed decisions based on comprehensive analyses of patient information [30].

AI systems can analyze data with a level of precision that reduces the likelihood of human error, leading to more reliable diagnoses and treatment plans [28, 30]. This increased accuracy is particularly vital in high-stakes environments such as oncology, where misdiagnosis can have severe consequences [29]. AI can enhance operational efficiency by automating routine tasks, such as data entry and appointment scheduling, which allows healthcare providers to allocate their time and resources more effectively [31–33]. Another key benefit of AI in healthcare is its potential to personalize care. By analyzing individual patient data, AI can tailor treatment recommendations to align with each patient's unique health profile, preferences, and circumstances [34]. This personalized approach not only improves patient satisfaction but also enhances adherence to treatment protocols, as patients are more likely to engage with care that is relevant to their specific needs [28]. AI can facilitate shared decision-making between patients and healthcare providers, promoting a collaborative approach to health management that empowers patients to take an active role in their care [35]. AI algorithms have been developed to predict patient outcomes in critical care settings, allowing healthcare teams to make proactive decisions that can significantly improve survival rates [36]. In oncology, AI systems are being used to analyze pathology slides and identify cancerous cells with high accuracy, thereby expediting the diagnostic process and enabling timely treatment initiation [28, 29]. AI-driven tools have been implemented in mental health care to provide personalized interventions based on real-time patient data, enhancing the effectiveness of therapeutic approaches [23, 31].

Despite the numerous advantages of AI in healthcare, challenges remain in its widespread adoption. Concerns about data privacy, algorithmic bias, and the potential for diminished clinician-patient interactions must be addressed to ensure that AI technologies are implemented ethically and effectively [37–39]. Engaging healthcare professionals and patients in discussions about the role of AI in care delivery is essential for promoting trust and acceptance of these innovations [40]. Ongoing education and training for healthcare providers are crucial to equip them with the skills necessary to effectively utilize AI tools in their practice [21, 41]. AI is poised to revolutionize healthcare by enhancing diagnostics, treatment, and patient care through improved accuracy, efficiency, and personalization. The successful integration of AI-driven tools into healthcare systems has the potential to significantly improve health outcomes and patient satisfaction. However, addressing ethical concerns and ensuring proper training for healthcare professionals will be vital for the successful adoption of AI technologies in clinical practice.

3 Applications of AI in Healthcare

3.1 *AI in Diagnostics*

One of AI's most transformative applications is in diagnostics, particularly through medical imaging and predictive analytics. ML algorithms, especially deep learning models, can analyze complex imaging data with a level of accuracy often comparable to expert clinicians. CNNs have demonstrated substantial success in interpreting radiological images, such as identifying abnormalities in X-rays, CT scans, and MRIs. A study by Esteva et al. [8] demonstrated that CNNs could classify skin cancer with diagnostic accuracy on par with dermatologists, a breakthrough in early-stage skin cancer detection. Similarly, deep learning models have been trained to detect lung nodules in CT scans, significantly aiding in early diagnosis and treatment planning for lung cancer. Beyond imaging, AI algorithms are also being developed for predictive diagnostics, which leverage large datasets from EHRs to anticipate patient outcomes. By identifying patterns in patient histories and genetic information, predictive models can assess risk factors and foresee potential complications, enabling proactive healthcare decisions. For example, studies show that AI models can predict sepsis—a life-threatening infection that is challenging to detect early—with a high degree of accuracy [42]. This allows clinicians to intervene sooner, improving patient survival rates. NLP has further enhanced diagnostics by extracting valuable insights from unstructured clinical data, such as doctor's notes and patient histories. NLP-based algorithms can mine these records to identify potential health risks, suggest diagnoses, and flag anomalies for further examination [29]. For instance, NLP tools have been effectively used to automate the identification of pneumonia cases from radiology reports, allowing for quicker triage and treatment [43].

3.2 *AI in Treatment Planning*

AI's role in treatment planning is primarily focused on optimizing therapeutic decisions based on patient-specific data. Machine learning models, particularly those leveraging vast clinical datasets, can analyze individual health metrics and suggest personalized treatment plans tailored to a patient's unique profile. This approach has seen promising applications in oncology, where precision medicine is crucial. IBM's Watson for Oncology, for example, uses AI to recommend cancer treatment options by comparing individual patient data with a comprehensive database of medical literature, clinical trials, and patient histories. Studies have shown that Watson's recommendations align with expert oncologists' choices in a substantial number of cases, underscoring its potential to support clinical decision-making [44]. AI has revolutionized pharmacology by aiding in drug discovery and repurposing. Traditional drug development can take years and substantial financial investment. However, AI algorithms can rapidly analyze large datasets to identify potential drug candidates, reducing both time and cost. For example, BenevolentAI has developed AI-driven solutions to accelerate drug discovery, successfully identifying existing drugs with potential applications for new diseases [45, 46]. This was evidenced in the early phases of the COVID-19 pandemic when AI was used to screen approved drugs for efficacy against the virus, contributing to the accelerated development of treatment options [47]. AI is also being used to develop personalized treatment regimens for chronic diseases, such as diabetes and heart disease [48, 49]. Wearable devices and mobile applications can monitor patient health metrics in real time, such as blood sugar levels, blood pressure, and heart rate [45, 50]. AI algorithms analyze this data to recommend lifestyle adjustments, medications, or interventions, allowing patients to manage their conditions more effectively and reducing the risk of complications [51].

3.3 AI in Patient Care and Monitoring

AI applications in patient care extend beyond diagnostics and treatment into continuous monitoring and post-treatment care. Remote patient monitoring (RPM) systems, powered by AI and connected devices, have proven highly effective in managing chronic conditions and monitoring post-surgical recovery [52]. For example, AI-enabled wearable devices can continuously track vital signs and send real-time updates to healthcare providers, alerting them to any significant changes [53]. This is particularly valuable in managing high-risk patients who require consistent monitoring but may not need immediate hospitalization, thus reducing healthcare costs and resource burdens [54]. In cardiac care, AI is used to predict and prevent heart-related events. For instance, AI models can analyze ECG data from wearable devices to detect arrhythmias and other cardiac anomalies early, often before patients experience any symptoms. This early detection allows for timely interventions, which can prevent severe cardiac events and improve long-term outcomes. Atrial fibrillation, a common but often undetected heart rhythm disorder, has been effectively monitored through AI algorithms applied to wearable device data, demonstrating AI's potential to enhance preventive cardiac care [55]. In mental health, AI has shown promise in identifying early signs of mental health issues, such as depression and anxiety [56, 57]. Chatbots and virtual assistants powered by NLP can engage patients in conversation, analyze speech patterns and sentiment, and assess potential mental health risks [58]. These tools can then provide initial support, guide users to resources, or recommend that patients seek professional help. Research indicates that AI-driven mental health apps can support early intervention strategies, providing a valuable addition to traditional mental health care [59].

3.4 Examples of Successful AI Applications in Healthcare

Several successful AI applications underscore the potential of AI in healthcare. IBM's Watson for Oncology has been widely studied and used in clinical settings to support oncologists in recommending treatment plans [60, 61]. By reviewing extensive clinical literature and patient data, Watson aids in identifying effective therapies and aligning recommendations with current medical research, thus supporting informed clinical decisions [44]. In radiology, Zebra Medical Vision uses AI to analyze imaging data for various conditions, including liver disease, lung cancer, and osteoporosis [62]. The company's AI algorithms have demonstrated high accuracy in identifying pathologies and have been implemented in multiple healthcare systems to support radiologists [9]. AI has also made strides in improving patient engagement and monitoring through digital health platforms. For example, Apple's HealthKit and Google Fit use AI to monitor health metrics, analyze patterns, and provide insights into users' daily health behaviors [63]. These applications promote proactive health management, encouraging individuals to adopt healthier lifestyles and adhere to prescribed treatments. The application of AI in healthcare has led to significant advancements across diagnostics, treatment, and patient care, providing tools that enhance accuracy, efficiency, and personalization.

4 Regulatory Frameworks and Standards

4.1 Overview of Existing Regulations Relevant to AI in Healthcare

Two primary regulatory frameworks are pivotal in guiding AI usage in healthcare: the Health Insurance Portability and Accountability Act (HIPAA) in the United States and the General Data Protection Regulation (GDPR) in the European Union [64, 65]. These frameworks focus on protecting individual data, ensuring consent, and safeguarding against unauthorized data use, forming the bedrock of patient privacy protection. HIPAA, established in 1996, regulates the use, storage, and sharing of protected health information (PHI). HIPAA mandates that healthcare providers, insurers, and other stakeholders implement strong security measures for handling sensitive patient data. In the context of AI, HIPAA applies to the storage and use of PHI in machine learning systems, requiring data encryption, secure transmission, and data anonymization when feasible. HIPAA's Privacy and Security Rules outline guidelines for ensuring data confidentiality and integrity, critical when handling data-intensive AI applications [64].

GDPR, enacted in 2018, serves as a comprehensive data privacy law in the EU and extends to all entities processing data from EU residents, including healthcare providers and AI technology developers [65]. GDPR emphasizes the need for explicit, informed consent and grants individuals the right to access, rectify, and erase their data. Importantly, GDPR imposes strict conditions on automated decision-making, particularly when algorithms use personal data for health predictions [66]. With AI, GDPR's emphasis on transparency and individual rights is crucial, as it requires AI systems to be explainable and to limit personal data processing to legitimate, necessary purposes [66]. Other notable frameworks include the California Consumer Privacy Act (CCPA), which mirrors GDPR principles but applies in the United States, and the International Medical Device Regulators Forum (IMDRF) guidelines, which address regulatory requirements for AI as a medical device [67]. These regulations set standards for responsible AI use and ensure that developers and healthcare providers adhere to established protocols for data security and patient privacy.

4.2 Role of Regulatory Bodies in Ensuring Security and Privacy

Regulatory bodies play a critical role in enforcing data protection standards, overseeing healthcare providers, and ensuring that AI applications meet stringent security and privacy standards. For instance, the U.S. Department of Health and Human Services enforces HIPAA compliance through regular audits, fines for breaches, and guidance for healthcare providers on maintaining data privacy [68]. Similarly, the European Data Protection Board (EDPB) monitors GDPR compliance, issuing fines for violations and establishing guidelines for data handling and automated decision-making [69]. Regulatory bodies also collaborate with industry experts, policymakers, and researchers to adapt regulations to new technologies. By working closely with AI and healthcare experts, these agencies stay informed on AI advancements and can update privacy standards and frameworks to cover new risks. Through this proactive approach, regulatory bodies help maintain public trust in AI systems and establish accountability measures that protect sensitive healthcare data.

4.3 Challenges in Adapting Regulations to Evolving AI Technologies

AI's rapid evolution presents significant challenges for regulatory frameworks, as existing laws may not address the specific risks posed by advanced AI systems. For example, HIPAA and GDPR were not designed with the complexities of machine learning algorithms in mind, creating gaps in regulatory coverage [70]. AI systems that rely on large datasets may require more flexible interpretations of data minimization and consent requirements, which traditional frameworks do not fully address. One notable challenge is the "black box" nature of some AI models, particularly deep learning algorithms that operate through complex, non-linear processes that are difficult to interpret [71]. Regulations like GDPR mandate transparency but ensuring that AI systems are fully explainable is challenging and may not be feasible with all models. This lack of explainability creates regulatory hurdles and complicates efforts to hold AI systems accountable for their decisions. Regulatory adaptation must keep pace with AI's cross-border applications. Since healthcare and AI technologies operate globally, discrepancies in regulatory frameworks across jurisdictions can hinder the consistent application of privacy standards. Harmonizing global standards is critical but challenging, requiring international cooperation and ongoing dialogue among policymakers, regulatory agencies, and technology developers.

5 Strategies for Enhancing Security and Privacy in Healthcare AI

5.1 Data Encryption, Anonymization, and Secure Storage Practices

To protect sensitive patient data, healthcare providers can employ encryption techniques to secure data both in transit and at rest. Encryption ensures that data is unreadable without the proper decryption key, protecting it from unauthorized access even if a breach occurs. Additionally, anonymizing data by removing personally identifiable information (PII) can enhance security by reducing the risk of patient re-identification in datasets used for AI training and analysis. Secure storage practices are equally important, with cloud providers offering healthcare-specific services that comply with regulations like HIPAA and GDPR. By implementing multi-layered security measures, such as firewalls and access controls, healthcare organizations can safeguard data while maintaining accessibility for authorized users.

5.2 Privacy-Preserving Machine Learning Techniques

Privacy-preserving machine learning techniques, such as federated learning and differential privacy, offer solutions to data privacy concerns in AI [72, 73]. Federated learning enables multiple institutions to train AI models collaboratively without centralizing their data, reducing the risk of data breaches [73]. Instead of transferring data to a central location, federated learning keeps data on local devices and shares only aggregated model updates, preserving privacy and minimizing data exposure. Differential privacy adds controlled "noise" to data, preventing specific patient information from being identifiable while still allowing meaningful patterns to emerge [72]. This technique has proven effective in maintaining privacy for machine learning applications and is gaining traction as a standard practice in privacy-focused AI.

5.3 Implementing AI Ethics Guidelines and Privacy-Focused Policies

Healthcare providers can implement ethics guidelines and policies specifically tailored to AI usage, ensuring that AI systems prioritize patient rights and transparency. Ethical guidelines, such as those proposed by the AI4People initiative and the IEEE's Ethically Aligned Design framework, provide standards for responsible AI use, including fairness, accountability, and explainability [12, 74]. By adopting these guidelines, healthcare providers can ensure that their AI systems align with ethical principles and respect patient privacy. Establishing privacy-focused policies, such as consent procedures for data collection and transparency measures for AI-driven decision-making, can further enhance security and patient trust. Consent forms that inform patients about AI's role in their care, the data collection process, and their rights offer a foundation for ethical AI deployment.

6 Future Trends and Directions

Several emerging technologies hold promises for improving AI security and privacy. Blockchain, for example, is increasingly explored to secure patient data by creating decentralized, tamper-resistant ledgers. Blockchain enables secure data sharing among healthcare providers and patients, ensuring data integrity and traceability. Quantum cryptography also offers potential advancements in encryption, providing stronger protection against cyber-attacks as quantum computing becomes more widespread. AI has the potential to enhance its own security by using predictive analytics to anticipate and mitigate cyber threats. Machine learning algorithms can identify suspicious behavior patterns and detect anomalies, enabling healthcare providers to respond to potential security breaches before they escalate. For instance, AI systems can be trained to recognize early signs of ransomware attacks, such as unusual data access patterns, and isolate affected systems to prevent further spread. As AI technology continues to evolve, several challenges and ethical considerations will need to be addressed. These include refining AI transparency, preventing algorithmic bias, and ensuring that AI systems uphold patient rights across diverse populations. Another key challenge will be developing adaptive regulatory frameworks that can keep pace with rapid AI advancements without stifling innovation. Ethically, AI developers must prioritize designing systems that safeguard human rights and dignity. This includes addressing biases within AI models, enhancing algorithm transparency, and ensuring that AI systems support healthcare providers rather than replacing human judgment. Ongoing collaboration between regulatory bodies, technology developers, and healthcare professionals will be essential in promoting responsible and secure AI applications.

7 Conclusion

The integration of AI in healthcare offers transformative potential, enhancing diagnostics, personalizing treatments, and optimizing operational efficiency. AI technologies, from machine learning to natural language processing, have already demonstrated their value in improving patient outcomes and streamlining clinical workflows. However, this advancement brings a unique set of challenges, especially concerning data security, patient privacy, and ethical considerations. Compliance with regulatory frameworks like HIPAA and GDPR is crucial to ensure that AI applications safeguard patient rights and maintain data confidentiality. Privacy-preserving techniques such as federated learning and differential privacy provide innovative solutions to address these challenges, enabling the continued development of AI in a secure and ethical manner. Moving forward, collaboration among regulatory bodies, technology developers, and healthcare professionals will be essential to create adaptive policies that balance innovation with accountability. As new technologies such as blockchain and quantum cryptography emerge, they will play a pivotal role in fortifying AI applications against privacy risks. This chapter underscores the importance of a cautious yet progressive approach to AI in healthcare, advocating for responsible innovation that prioritizes both technological advancement and the trust of healthcare stakeholders.

References

1. Qayyum, M.U., et al.: Revolutionizing healthcare: the transformative impact of artificial intelligence in medicine. BIN: Bull. Inf. **1**(2), 71–83 (2023)
2. Govindaraj, M., et al.: Revolutionizing healthcare: the transformative impact of artificial intelligence. In: Revolutionizing the Healthcare Sector with AI, pp. 54–78. IGI Global (2024)
3. Van Leeuwen, K.G., et al.: How does artificial intelligence in radiology improve efficiency and health outcomes? Pediatric Radiol. 1–7 (2022)
4. Esteva, A., et al.: Deep learning-enabled medical computer vision. NPJ Digital Med. **4**(1), 5 (2021)
5. Winter, P.D., Carusi, A.: (De) troubling transparency: artificial intelligence (AI) for clinical applications. Med. Humanit. **49**(1), 17–26 (2023)
6. Sharma, P., et al.: Artificial intelligence in diagnostic imaging: status quo, challenges, and future opportunities. J. Thorac. Imaging **35**, S11–S16 (2020)
7. Tadiboina, S.N.: The use of AI in advanced medical imaging. J. Positive School Psychol. **6**(11), 1939–1946 (2022)
8. Esteva, A., et al.: Dermatologist-level classification of skin cancer with deep neural networks. Nature. **542**(7639), 115–118 (2017)
9. Lakhani, P., Sundaram, B.: Deep learning at chest radiography: automated classification of pulmonary tuberculosis by using convolutional neural networks. Radiology **284**(2), 574–582 (2017)
10. Jiang, F., et al.: Artificial intelligence in healthcare: past, present and future. Stroke Vascular Neurol. **2**(4) (2017)
11. Rajkomar, A., et al.: Machine learning in medicine. N. Engl. J. Med. **380**(14), 1347–1358 (2019)

12. Shah, N.H., et al.: Making machine learning models clinically useful. JAMA **322**(14), 1351–1352 (2019)
13. Finlayson, S.G., et al.: Adversarial attacks on medical machine learning. Science **363**(6433), 1287–1289 (2019)
14. Hirschberg, J., Manning, C.D.: Advances in natural language processing. Science **349**(6245), 261–266 (2015)
15. Devapatla, H., Katti, S.R.: Streamlining administrative processes in healthcare through robotic process automation: a comprehensive examination of RPA's impact on billing, scheduling, and claims processing. Afr. J. Artif. Intell. Sustain. Dev. **3**(2), 14–27 (2023)
16. Acemoglu, D., Restrepo, P.: Artificial intelligence, automation, and work. In: The Economics of Artificial Intelligence: An Agenda, pp. 197–236. University of Chicago Press (2018)
17. Barrett, M., et al.: Artificial intelligence supported patient self-care in chronic heart failure: a paradigm shift from reactive to predictive, preventive and personalised care. EPMA J. **10**, 445–464 (2019)
18. Scott, I.A., et al.: Exploring stakeholder attitudes towards AI in clinical practice. BMJ Health Care Inf. **28**(1) (2021)
19. Yin, J., et al.: Role of artificial intelligence applications in real-life clinical practice: systematic review. J. Med. Internet Res. **23**(4), e25759 (2021)
20. Emani, S., et al.: Physicians' perceptions of and satisfaction with artificial intelligence in cancer treatment: a clinical decision support system experience and implications for low-middle–income countries. JMIR Cancer **8**(2), e31461 (2022)
21. Russell, R.G., et al.: Competencies for the use of artificial intelligence–based tools by health care professionals. Acad. Med. **98**(3), 348–356 (2023)
22. Castagno, S., Khalifa, M.: Perceptions of artificial intelligence among healthcare staff: a qualitative survey study. Front. Artif. Intell. **3**, 578983 (2020)
23. Esmaeilzadeh, P., et al.: Patients' perceptions toward human–artificial intelligence interaction in health care: experimental study. J. Med. Internet Res. **23**(11), e25856 (2021)
24. Chee, M.L., et al.: Artificial intelligence applications for COVID-19 in intensive care and emergency settings: a systematic review. Int. J. Environ. Res. Public Health **18**(9), 4749 (2021)
25. Mann, S., et al.: Artificial intelligence applications used in the clinical response to COVID-19: A scoping review. PLOS Digital Health **1**(10), e0000132 (2022)
26. Rasool, S., et al., Innovations in AI-powered healthcare: transforming cancer treatment with innovative methods. BULLET: Jurnal Multidisiplin Ilmu **3**(1), 118–128 (2024)
27. Bekbolatova, M., et al.: Transformative potential of AI in healthcare: definitions, applications, and navigating the ethical Landscape and Public perspectives. Healthcare. MDPI (2024)
28. Haleem, A., et al.: Current status and applications of Artificial Intelligence (AI) in medical field: An overview. Curr. Med. Res. Practice **9**(6), 231–237 (2019)
29. Istasy, P., et al.: The impact of artificial intelligence on health equity in oncology: scoping review. J. Med. Internet Res. **24**(11), e39748 (2022)
30. Lee, D., Yoon, S.N.: Application of artificial intelligence-based technologies in the healthcare industry: opportunities and challenges. Int. J. Environ. Res. Public Health **18**(1), 271 (2021)
31. Zidaru, T., et al.: Ensuring patient and public involvement in the transition to AI-assisted mental health care: a systematic scoping review and agenda for design justice. Health Expect. **24**(4), 1072–1124 (2021)
32. Pradhan, K., et al.; Use of artificial intelligence in healthcare delivery in India. J. Hospital Manage. Health Policy. **5** (2021)
33. Tran, V.-T., et al.: Patients' views of wearable devices and AI in healthcare: findings from the ComPaRe e-cohort. NPJ Digital Med. **2**(1), 53 (2019)
34. Chew, H.S.J., Achananuparp, P.: Perceptions and needs of artificial intelligence in health care to increase adoption: scoping review. J. Med. Internet Res. **24**(1), e32939 (2022)
35. Witkowski, K., et al.: Public perceptions of artificial intelligence in healthcare: ethical concerns and opportunities for patient-centered care. BMC Med. Ethics **25**(1), 1–11 (2024)
36. Santos, D., et al.: Patterns and outcomes of intensive care on acute ischemic stroke patients in the US. Circ.: Cardiovascular Qual. Outcomes **16**(3), e008961 (2023)

37. Murphy, K., et al.: Artificial intelligence for good health: a scoping review of the ethics literature. BMC Med. Ethics **22**, 1–17 (2021)
38. Richardson, J.P., et al.: Patient apprehensions about the use of artificial intelligence in healthcare. NPJ Digital Med. **4**(1), 140 (2021)
39. Nitiéma, P.: Artificial intelligence in medicine: text mining of health care workers' Opinions. J. Med. Internet Res. **25**, e41138 (2023)
40. Jeyakumar, T., et al.: Preparing for an artificial intelligence–enabled future: patient perspectives on engagement and health care professional training for adopting artificial intelligence technologies in health care settings. JMIR AI **2**(1), e40973 (2023)
41. Abdullah, R., Fakieh, B.: Health care employees' perceptions of the use of artificial intelligence applications: survey study. J. Med. Internet Res. **22**(5), e17620 (2020)
42. Shashikumar, S.P., et al.: DeepAISE—An end-to-end development and deployment of a recurrent neural survival model for early prediction of sepsis. arXiv:1908.04759 (2019)
43. Wang, X., et al.: Chestx-ray8: hospital-scale chest x-ray database and benchmarks on weakly-supervised classification and localization of common thorax diseases. In: Proceedings of the IEEE Conference on Computer Vision and Pattern Recognition (2017)
44. Somashekhar, S., et al.: Watson for oncology and breast cancer treatment recommendations: agreement with an expert multidisciplinary tumor board. Ann. Oncol. **29**(2), 418–423 (2018)
45. Shanmuga Sundari, M., et al.: AI-based personalized drug treatment. In: Artificial Intelligence and Machine Learning in Drug Design and Development, pp. 369–406 (2024)
46. Khanna, A., Jain, S.: Personalized drug treatment: transforming healthcare with AI. In: Artificial Intelligence and Machine Learning in Drug Design and Development, pp. 295–319 (2024)
47. Pushpakom, S., et al.: Drug repurposing: progress, challenges and recommendations. Nat. Rev. Drug Discovery **18**(1), 41–58 (2019)
48. Rahman, K., et al.: Leveraging AI for chronic disease management: a new horizon in medical research. Malaysian J. Med. Biol. Res. **9**(2), 81–90 (2022)
49. Abdullah, M.: Integrating artificial intelligence in precision medicine: enhancing predictive accuracy and personalized care in chronic disease management. Med. Life Sci. **3**(1), 6–15 (2024)
50. Knight, S.R., et al.: Mobile devices and wearable technology for measuring patient outcomes after surgery: a systematic review. NPJ Digital Med. **4**(1), 157 (2021)
51. Martinez-Rodrigo, A., et al.: Development of a recommendation system and data analysis in personalized medicine: an approach towards healthy vascular ageing. Health Inf. Sci. Syst. **12**(1), 1–16 (2024)
52. Murugan, R., et al.: Amalgamation of transfer learning and explainable AI for internet of medical things. Recent Adv. Comput. Sci. Commun. (Formerly: Recent Patents Comput. Sci.). **17**(4), 40–53 (2024).
53. Chawla, N., Dalal, S.: Edge AI with wearable IoT: a review on leveraging edge intelligence in wearables for smart healthcare. In: Green Internet of Things for Smart Cities, pp. 205–231 (2021)
54. Krittanawong, C., et al.: Artificial intelligence in precision cardiovascular medicine. J. Am. Coll. Cardiol. **69**(21), 2657–2664 (2017)
55. Rajpurkar, P., et al.: Deep learning for chest radiograph diagnosis: a retrospective comparison of the CheXNeXt algorithm to practicing radiologists. PLoS Med. **15**(11), e1002686 (2018)
56. Zafar, F., et al.: The role of artificial intelligence in identifying depression and anxiety: a comprehensive literature review. Cureus **16**(3) (2024)
57. Bickman, L.: Improving mental health services: A 50-year journey from randomized experiments to artificial intelligence and precision mental health. Admin. Policy Mental Health Mental Health Services Res. **47**(5), 795–843 (2020)
58. Russo, A., et al.: Dialogue systems and conversational agents for patients with dementia: the human–robot interaction. Rejuvenation Res. **22**(2), 109–120 (2019)
59. Fitzpatrick, K.K., et al.: Delivering cognitive behavior therapy to young adults with symptoms of depression and anxiety using a fully automated conversational agent (Woebot): a randomized controlled trial. JMIR Mental Health **4**(2), e7785 (2017)

60. Tupasela, A., Di Nucci, E.: Concordance as evidence in the Watson for oncology decision-support system. AI Soc. **35**, 811–818 (2020)
61. Zhou, N., et al.: Concordance study between IBM Watson for oncology and clinical practice for patients with cancer in China. Oncologist **24**(6), 812–819 (2019)
62. Mintz, Y., Brodie, R.: Introduction to artificial intelligence in medicine. Minim. Invasive Ther. Allied Technol. **28**(2), 73–81 (2019)
63. Ballinger, B., et al.: DeepHeart: semi-supervised sequence learning for cardiovascular risk prediction. In: Proceedings of the AAAI Conference on Artificial Intelligence (2018)
64. Oakley, A.: HIPAA, HIPPA, or HIPPO: what really is the heath insurance portability and accountability act? Biotechnol. Law Rep. **42**(6), 306–318 (2023)
65. Bakare, S.S., et al.: Data privacy laws and compliance: a comparative review of the EU GDPR and USA regulations. Comput. Sci. IT Res. J. **5**(3), 528–543 (2024)
66. Brkan, M.: Do algorithms rule the world? Algorithmic decision-making and data protection in the framework of the GDPR and beyond. Int. J. Law Inf. Technol. **27**(2), 91–121 (2019)
67. Coravos, A., et al.: Digital medicine: a primer on measurement. Digital Biomarkers **3**(2), 31–71 (2019)
68. Choi, Y.B., Williams, C.E.: A HIPAA security and privacy compliance audit and risk assessment mitigation approach. In: Research Anthology on Securing Medical Systems and Records, pp. 706–725. IGI Global (2022)
69. Barrett, C.: Emerging trends from the first year of EU GDPR enforcement. Scitech Lawyer **16**(3), 22–35 (2020)
70. Forcier, M.B., et al.: Integrating artificial intelligence into health care through data access: can the GDPR act as a beacon for policymakers? J. Law Biosci. **6**(1), 317–335 (2019)
71. Hassija, V., et al.: Interpreting black-box models: a review on explainable artificial intelligence. Cogn. Comput. **16**(1), 45–74 (2024)
72. Dodda, S.B., et al.: Federated learning for privacy-preserving collaborative AI: exploring federated learning techniques for training AI models collaboratively while preserving data privacy. Aust. J. Mach. Learn. Res. Appl. **2**(1), 13–23 (2022)
73. Rodríguez-Barroso, N., et al.: Federated learning and differential privacy: software tools analysis, the Sherpa. ai FL framework and methodological guidelines for preserving data privacy. Inf. Fusion **64**, 270–292 (2020)
74. Floridi, L., et al.: AI4People—An ethical framework for a good AI society: opportunities, risks, principles, and recommendations. Mind. Mach. **28**, 689–707 (2018)

Eriona Çela is a distinguished academic and globally recognized expert in artificial intelligence, machine learning, deep learning, and natural language processing, alongside her noted contributions to European integration and negotiation. Renowned for her pioneering research in AI technologies, Dr. Çela has advanced the application of AI and machine learning in fields such as public administration, education, and policy development, bringing innovative solutions to complex social and organizational challenges. She holds prominent positions within several key institutions, including the Academy of European Integration and Negotiations (AIEN), the Albanian School of Public Administration (ASPA), and the Business Academy College, where her influence extends across multiple disciplines. As a Jean Monnet Chair holder at the Institute of European Studies, University of Tirana, Dr. Çela has led major projects, including the European Integration School 2023's eighth edition, and has earned acclaim for her role in shaping EU policies and enlargement strategies through AI-driven insights. Her research is widely published in prestigious journals such as the European Journal of Arts, Humanities and Social Sciences, the European Journal of Contemporary Education, and the International Journal of Balkan Studies. Dr. Çela is also an Associate Editor for the International Journal of Risk and Contingency Management (IJRCM), an IGI Journal, and is actively involved with international academic institutions, including the OSCE Academy in Bishkek, Vytautas Magnus University, and Università degli Studi di Bari Aldo Moro, fostering collaborative AI research on a European scale. Dr. Çela's contributions to educational leadership are highlighted through her roles with Shkolla e Drejtorëve (CSL) and the AADF, where she focuses on developing future leaders equipped with cutting-edge AI competencies. Recognized with numerous accolades for her contributions to both AI research and public administration, Dr. Çela is a sought-after keynote speaker at global conferences and policy-making forums. Her work continues to shape the future of AI in education and public administration in Albania and beyond, setting a benchmark for integrating advanced AI technologies with European educational standards and practices.

Mathias Fonkam is an accomplished Associate Teaching Professor at Penn State University with over 20 years of experience in computer science education and as a practicing software engineer. He holds a Ph.D. in Computer Science, an M.Sc. in Systems Engineering and a B.Sc. in Computer Engineering from Cardiff University, UK. His professional journey includes serving as the Dean at the American University of Nigeria (AUN), where he spearheaded numerous initiatives, such as developing the university's Open-Source ERP System and establishing various postgraduate programs. Dr. Fonkam's research interests span AI, data science, blockchain, open-source software and systems dynamics, with over 30 publications in reputable conferences and journals. He has also contributed significantly to curriculum development and accreditation processes in Nigerian universities. Fluent in English, Portuguese, and French, Dr. Fonkam is dedicated to advancing computing education and professional development, leveraging his extensive expertise in teaching, mentoring, and research.

Alexey Vedishchev is an experienced IT professional and educator with over 12 years of expertise in computer science education. He holds an MSc in Information Technology and Business from Altay State Technical University in Russia, where he also completed specialized training in information management for higher education. Since 2014, he has been an instructor at the American University of Nigeria, where he teaches various programming, database, and software development courses. Alexey's teaching style emphasizes research, project-based learning, and diverse instructional methods. His professional background includes roles in IT support and management at organizations like Russian Post and Quanteq LTD, focusing on product support, IT policy development, and infrastructure planning.

Narasimha Rao Vajjhala is an academic and researcher currently serving as Professor and Chair of the Computer Science Department at the American University in Bulgaria (AUBG). He is also the Principal Investigator for the Democracy Project (DemPro) at the Center for Information, Democracy, and Citizenship (CIDC) at AUBG. Dr. Vajjhala previously held leadership roles as Dean of the Faculty of Engineering and Architecture at the University of New York Tirana (UNYT) in Albania and Chair of the Computer Science and Software Engineering programs at the American University of Nigeria (AUN). With over 23 years of experience in academia, he has taught programming, database management, and related courses across Europe and Africa at both undergraduate and graduate levels. An active contributor to the global research community, Dr. Vajjhala is a Senior Member of ACM and IEEE, Editor-in-Chief of the International Journal of Risk and Contingency Management (IJRCM), and a member of the Project Management Institute (PMI). Beyond academia, he has worked as a technology consultant for European firms and has been actively involved in EU-funded projects. Dr. Vajjhala holds a Doctorate in Information Systems and Technology from the United States, a Master of Science in Computer Science and Applications from India, and a Master of Business Administration (MBA) specializing in Information Systems from Switzerland. His expertise spans computer science, project management, and information systems, making significant contributions to both research and industry.

AI in Diagnostic and Disease Prediction

Hermehar P. S. Bedi, Renu Puri, Amandeep Kaur, and Deepak Puri

Abstract This chapter examines the essential role of early cancer detection in reducing mortality rates amidst the rising global burden of the disease. According to the World Health Organization (WHO), cancer claimed 10 million lives in 2020, underscoring its position as a leading cause of death worldwide. In India, data from the National Cancer Registry Program (ICMR) revealed a concerning increase in cancer cases, growing from 1.3 million in 2020 to 1.4 million in 2022, accompanied by a corresponding rise in mortality rates. Traditional diagnostic tools, such as Magnetic Resonance Imaging (MRI), Positron Emission Tomography (PET), and Computerized Tomography (CT), depend on expert interpretation, which is often time-consuming and prone to human error, potentially delaying diagnosis. To address these challenges, this chapter explores the transformative role of Artificial Intelligence (AI) and Machine Learning (ML) in enhancing cancer detection through automated imaging analysis. Leveraging advancements in AI-ML, innovative techniques and models have emerged, allowing for faster, more accurate image analysis and reducing reliance on manual review. The chapter provides an in-depth analysis of the most impactful AI-ML techniques, emphasizing their ability to improve diagnostic accuracy and efficiency. Additionally, a comprehensive literature review offers insights into how Machine Learning and Deep Learning (DL) are revolutionizing cancer diagnostics, enabling more reliable, swift, and precise detection methods.

Keywords Cancer detection · Artificial intelligence · Machine learning · Early diagnosis · Medical imaging · Deep learning · MRI scans · PET scans · Automated analysis · Cancer diagnostics

---

H. P. S. Bedi (✉)
Department of Human-Centered Design, Indraprastha Institute of Information & Technology, Delhi, India
e-mail: hermeharbedi@gmail.com

R. Puri · A. Kaur
Department of Computer Science & Engineering, Chandigarh Engineering College—CGC Landran, Mohali, Punjab, India

D. Puri
Doaba College of Pharmacy, Mohali, Punjab, India

N. R. Vajjhala et al. (eds.), *Artificial Intelligence in Healthcare Information Systems—Security and Privacy Challenges*, Information Systems Engineering and Management 34, https://doi.org/10.1007/978-3-031-84404-1_2

1 Introduction

Machine learning is used in many fields, including education and healthcare. The use of machine learning has been significantly boosted by the accessibility of datasets and increased computing capacity brought about by technological breakthroughs. Many disciplines of healthcare use machine learning. Imagery, patient records, and other varieties of information gathered by the healthcare industry can be used to recognize trends and make predictions. Several concerns in healthcare are tackled using machine learning. Every person's amount of heart disease is different because heart disease is a personalized disorder. Building a machine learning model, training it on the dataset, and adding specific patient data therefore can enhance prediction. The forecast estimate will be customized to that person utilizing the information entered. Avoiding type-2 diabetes can be accomplished by managing one's weight, way of life, and other factors. There is no known treatment for the illness caused by the coronavirus. China is where the coronavirus 2019 (COVID-19) was discovered. Although there are various treatments in use for it, there are no definite stages for treatment [1]. Medical diagnosis is a challenging and important process that needs to be completed quickly and correctly. Most of the time, a doctor's knowledge and experience are used to diagnose this. This may occasionally result in inaccurate results and high medical costs for patients' treatments. The global market for medical imaging is booming as more individuals are aware about how medical imaging technologies are being utilized to address a wide range of medical condition [2].

In the medical industry, artificial intelligence (AI) principally concentrates on building the techniques and algorithms used to discern if a system's performance in making diagnoses of diseases is realistic. A medical diagnostic specifies the ailment or conditions that generate a patient's disease and indications. The physician's physical examination and history are typically utilized to collect diagnostic information [3]. The domain of artificial intelligence (AI) is only one. Machine learning is the discipline of computer systems replicating artificial intellect. It combines statistics and machine learning, where computer vision focuses on problem-solving and problem identification while statistics emphasizes on machine learning, hypotheses, and other data.

In real life, clinicians rely upon the symptoms which sufferers describe their physical examinations, laboratory and radiological test reports etc. These all methods are used to estimate the patient's condition and to plan the next plan of action. However, little analytical focus has been devoted to the predictive power and accuracy that may be generated by just using laboratory data for the diagnosis of specific diseases. To generate predictive models based on DL and ML that could be utilized by doctors to think critically in the framework of a hospital, the goal of this work was to use laboratory data alone. Researchers then confirmed our model by comparing its predictions with professional diagnoses. To increase performance, we also generated an aggregation of DL and ML models. A disease refers to a condition that disrupts the normal functioning of cells, tissues, and organs. Diseases are commonly perceived as medical disorders with distinct signs and symptoms. When an individual is afflicted by an illness, they may exhibit various symptoms and signs, ranging in intensity from mild to severe, depending on the underlying medical issue. Consequently, understanding what is considered normal for an entity becomes essential in detecting different diseases, as distinguishing between an ill entity and a healthy one is not always straightforward.

Diseases typically have multiple causes rather than being attributed to a single cause. Eventually, we experience symptoms of a disorder, such as headaches, coughing, colds, or weakness. These indications are commonly referred to as "symptoms." Almost every disease begins to show symptoms shortly after a person becomes ill. The severity of the illness affects how noticeable these symptoms are. Diseases can be categorised into two types: communicable diseases, and non-communicable diseases.

2 Machine Learning

Machine Learning (ML) is the subfield of Artificial Intelligence (AI), through which the machines learn the unique specific patterns and try to predict the output based on these unique patterns. In other words, ML is the way, through which the machines can learn information, which is extracted from enough data being supplied and which is used for the training purposes. The efficiency of the algorithm or model depends upon the quantity and quality of the data being used. Further, ML is classified into three types as depicted in Fig. 1, [4].

2.1 Supervised Learning

Supervised learning, a type of machine learning includes observing the system while it is learning. Most prevalent example of this type is the process of categorising

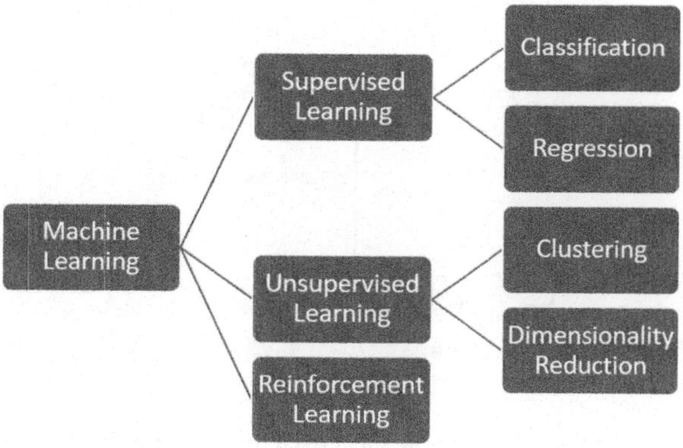

Fig. 1 Types of machine learning

objects into different classes. Supervised learning is further classified into two types as described in the Fig. 1. A set of n features A1, A2,..., An is often provided in supervised learning, which is assessed on n observations. A response B is then also measured on those same n observations. The next step is to use A1, A2,..., An to estimate B.

Using a labelled training dataset, the foundational procedure in supervised machine learning techniques is initially developed. The trained algorithm is then supplied the unlabeled test dataset to classify the data into relevant agencies. Figure. 2 exhibits how supervised machine learning algorithms classify whether an image is square or triangle using an abstract dataset for three diabetes patients supervised learning techniques make good alternatives for classification and regression problems. In classification issues, discrete output variables act as the primary output variable [5]. The divisions for this variable encompass "red" and "black," "diabetes" and "nondiabetic," among others. The corresponding output variable in regression or classification, such as evaluating a person's risk of developing cardiovascular disease, has a financial meaning. It provides a brief synopsis.

K Nearest Neighbor (KNN)

The KNN classification algorithm is a simple and early statistical learning method. The value "K" represents the number of closest neighbors utilized, that can be explicitly defined in the object function Object() or estimated using the given upper bound. As a result, similar situations lead to almost identical classifications, and when classifying a new instance, it is compared to each existing instance. When dealing with an unknown sample, the nearest neighbor approach searches for k training examples that are close to the unknown sample. Predictions for the test instance can be generated using two distinct methods that convert the distance into a weight based on their distances from multiple neighbors. The algorithm has several advantages, including analytical tractability and simplicity. It serves as an effective classifier and carries out well in disease prediction, especially in HD prediction, as it relies on a single occurrence. There are various methods available to calculate the distance between

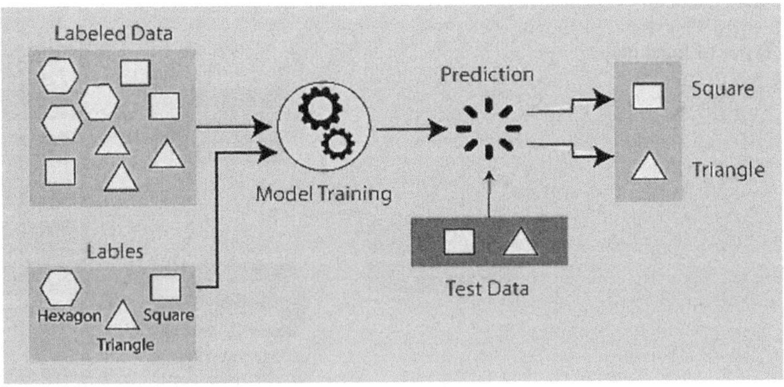

Fig. 2 Illustration of how supervised machine learning helps to categorize data from abstract data

instances and their nearest neighbors, with the most used being the Euclidean Formula (Eq. 1). Other frequently employed methods include the Manhattan and Minkowski Formulas (Eqs. 2 and 3).

$$\sqrt{\sum_{i=1}^{k} (x_i - y_i)^2} \quad (1)$$

$$\sum_{i=1}^{k} |x_i - y_i| \quad (2)$$

$$\left(\sum_{i=1}^{k} (|x_i - y_i|)^q\right)^{1/q} \quad (3)$$

This study found that setting the values of n neighbors to 2 and the leaf size to 40 resulted in the best fit for the dataset [6].

Logistic Regression

A classifier known as logistic regression can forecast binary outcomes from a set of independent factors. As a result, the target or dependent variable has a binary nature [7]. The target variable for logistic regression should be categorical in nature, according to the assumptions that should hold. The independent variables' multicollinearity was minimal or nonexistent. In the model, there should be a direct connection between the link function and the independent variable.

Random Forest (RF)

This method, also known as random choice forest, is an ensemble learning method, which is used for classification, regression, and other tasks. RF provides the statistical value of the individual trees such as mean prediction in case of regression and mode for the classification [8]. Random forests are particularly effective in mitigating the overfitting tendency of decision trees to their training set. Within the forest of trees, a discernible correlation was observed between the arrangement of trees and the likely outcomes.

Decision Tree (DT)

The realm of supervised learning algorithms includes Decision Tree calculation. Unlike its counterparts, this algorithm has the versatility to tackle both regression and classification tasks. The fundamental concept behind employing Decision Trees is to construct a training model that can predict classes or estimate target variables based on decision criteria learned from historical data (training data).

Support Vector Machine (SVM)

SVM, a supervised learning algorithm, finds application in both classification and regression tasks, although it is primarily employed for classification purposes. SVM is particularly effective in addressing various healthcare challenges and can handle both linear and non-linear problems. It employs a classification technique that seeks to create a linearly separable hyperplane to divide the dataset into two classes. Ultimately, the model can confidently predict the target labels for new instances. It constructs a hyperplane, which is based on a kernel function, which helps to divide the dataset into two distinct classes. This kernel function helps in creating two classes i.e., negative class and the positive class. Negative class is on the left side of the hyperplane and is generally denoted with -1, whereas positive class is on the right side of the hyperplane and is generally denoted with +1.

2.2 *Unsupervised Learning*

Unsupervised learning is also a subtype of ML, as discussed above. In this type of ML, the data provided is neither labelled nor classified and different algorithms are used to identify the patterns in these datasets. These algorithms identify the patterns in the dataset without anu supervision or human intervention [9]. It is again divided into two subgroups, as depicted in the Fig. 1. This kind of ML alludes to algorithms that examine influence in data sets that incorporate data points which are neither tagged nor categorized. As a corollary, the algorithms are unconstrained in their ability to classify, label, and reorganize the data points inside the data sets. It is not obligatory for users to keep an eye on the model. In other words, the learning algorithm is not equipped with labels, instead being left to figure out the organization in the input collection on its own. When no categories are available, unsupervised learning is when a software system combines unsorted data based on similarities and differences. The basic goal of unsupervised learning is to reveal unexplained patterns in unsupervised learning. Four components of unsupervised tasks are depicted in Fig. 3.

Clustering
The clustering technique groups or combines the items into clusters so that the items with the most similarities are placed together in one cluster whereas items with very less or no similarity are placed in different clusters. Data objects are categorized based on the commonalities identified through cluster analysis.

Association
Association rule learning is an unsupervised learning where the algorithm discovers patterns and relationships without a tutor because the data is unlabeled. The discovery of fascinating relationships hidden in massive datasets is often accomplished using association rules, which are descriptive rather than predictive. Usually, common item sets or rules are used to indicate the relationship [10]. Association rules mining is used to discover exciting new relationships between different objects in a set, common trends in transactional data, or any relational database. They are widely used in processes like price bundling, assortment selection, cross-selling, customer

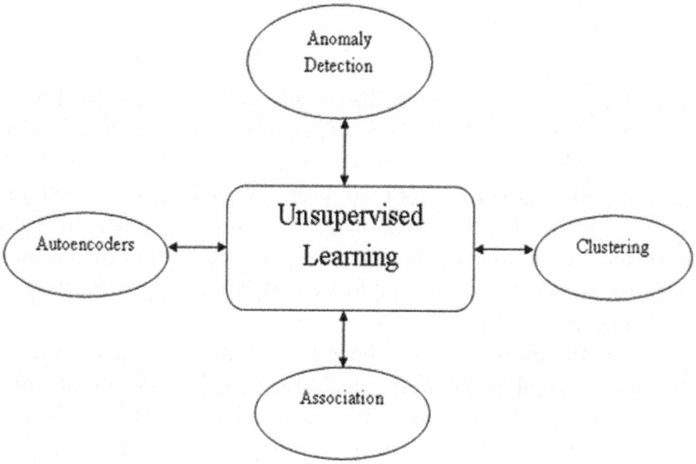

Fig. 3 Types of unsupervised learning

clustering in retail (which stores people tend to visit together), market basket analysis (which goods are purchased together), and other activities. This is an example of a what-if scenario that goes farther; if this, then that.

Anomaly Detection

Anomaly detection refers to any technique that finds outliers in a dataset. These anomalies can indicate incorrect network activity, reveal a malfunctioning sensor, or highlight data that must be cleaned up before analysis. An anomaly is when the data patterns deviate from or exceed the expected patterns. For instance, an odd traffic pattern on a network could be a sign that a compromised machine is sending sensitive information to an unapproved site. Anomalies are discovered by looking for or predicting data points that deviate from the anticipated pattern; if this, then that.

Autoencoders

The ability of neural networks to learn representations is used by autoencoders, an unsupervised learning technique. Feedforward neural networks with autoencoders have the same input and output. They first lower the dimension of the input and then reconstruct the output using this representation. The latent-space representation, often known as the code, effectively "summarizes" or "compresses" the input [10].

Reinforcement Learning

In this type of ML, a rewarding model is used for producing the desired results and punishes for all the undesired outcomes. It is most used in gaming but very less commonly also applied in other areas including the medical diagnosis and classification. Reinforcement learning (ML) is a way of analyzing tiny samples of data using mathematical and statistical methods to produce generalization.

3 Deep Learning

Deep learning (DL), a subgenre of ML uses multiple layers to extract both the high-level and low-level information from inputs such as images, numerical values and categorical values. Most of the modern DL models attribute Convolutional Neural Network (CNN), which can be combined with other DL models including embeddings, deep belief networks, and the Boltzmann machine. Three categories for DL reside: supervised, semi-supervised, and unsupervised. Some of the most popular kinds of designs are deep neural networks (DNN), reinforcement learning, and recurrent neural networks (RNN).

Different implementations for all these algorithms exist, including automated Through the use of several deep learning techniques on both communicable and non-communicable diseases, this research aims to identify relevant trends. Diseases are classified, after which many research projects on a given disease are published, along with their accuracy and findings. The findings of this study will aid the academics in better understanding the hotspots and trends in disease prediction models that use deep learning algorithms and in setting appropriate research objectives. In greater detail, this article only considered studies that employed multiple supervised machine learning algorithms in the same research environment for a single disease prediction. The main contribution of this work was made in this regard.

Deep learning (DL) is frequently applied in use cases such as text mining, spam detection, image classification and multimedia concept retrieval. In DL, the neural networks learn from the vast amount of data. Sometimes, DL is also referred to as Representation Learning [11]. Various deep learning techniques are employed for medical data analysis, encompassing CNN, encoded CNN, 3D CNN, U-Net, LSTM, RNN, ANN, and several others.

3.1 Long Short-Term Memory

LSTM, or Long Short-Term Memory, is a type of recurrent neural network (RNN) architecture designed to remember information for extended periods. It overcomes the limitations of traditional RNNs by using memory blocks to prevent issues like exploding and vanishing gradients, thereby enabling better handling of long-term dependencies in data.

The main disadvantage of CNN in grouping 3D MRI images was its low accuracy. This issue was addressed by LSTM, which provided better accuracy in classification. The model achieved an accuracy of 84% for classifying the glioma tumors in 3D MRI scans. This could even be improved by only adjusting hyperparameters such as the learning rate and employing data augmentation for better performance [12]. LSTM, an improved version of RNN, uses memory blocks instead of traditional RNN units to solve issues like exploding and vanishing gradients. The main difference between

LSTM and RNN is that LSTM maintains durable state by adding a cell state, enabling it to draw a connection between past data with present input [13].

Although deep CNNs have been used to identify brain tumors with good accuracy, their performance was less reliable for 3D medical data. An amalgamation of CNN and LSTM was employed to tackle this point, but it still lacked accuracy, indicating room for future development. LSTM performed exceptionally well in classifying 2D MRI images, achieving 98% accuracy when Gaussian filters were applied [12]. Shanthi et al. [14] proposed a hybrid model combining LSTM and CNN for automatic brain tumor classification, using the ARO algorithm to optimize weight values and enhance classification performance. This method achieved 97.5% accuracy by using CNN for feature map generation and LSTM for classification.

Another hybrid approach by Alsubai et al. [11] used CNN for feature extraction from cropped images. This model achieved remarkable results with 99.1% accuracy, 98.8% precision, 98.9% recall, and a 99% F1 measure, making it highly effective for brain MRI detection.

3.2 *Convolutional Neural Network*

DL utilizes Convolutional Neural Networks (CNN) as its most renowned and widely employed algorithm. One of the significant advantages of CNN differentiates to its predecessors in its capacity to automatically detect relevant attributes without the need of human intervention. CNNs have been widely used in various areas such as computer vision, speech processing, and Face Recognition [11]. As a multi-dimensional neural network, CNN is specifically designed for processing grid-structured data. It operates through a combination of two essential layers—the pooling layer and the convolution layer. The inspiration behind the architecture of CNNs draws from the neurons present in the human and animal brains, like conventional neural networks.

3.3 *Artificial Neural Network (ANN)*

It is a method that consists of a vast network of interconnected neurons, designed to learn from input data. When dealing with intricate medical data, this approach proves highly effective, as it efficiently learns and categorizes the complex information.

3.4 *Recurrent Neural Network (RNN)*

It is a kind of DL, in which the yield from the preceding stage is served as an input to the next step. In conventional neural networks, inputs and outputs are treated as independent entities. However, when predicting the next word in a sentence, the context of previous words becomes essential. This necessity arises from the need to retain information about the preceding words for accurate predictions. The primary and pivotal characteristic of RNN lies in its Hidden state, which retains crucial information about a sequence. This state is often known as the Memory State, as it stores the memory of previous inputs to the network. By employing shared weights across each element of the sequence, Recurrent Neural Networks (RNNs) reduce the parameter count and enable the model to adapt effectively to sequences of different lengths. This design choice not only promotes generalization to sequential data but also extends RNNs' capability to encompass other structured data types, including geographical or graphical data.

4 ML Models Used for Disease Prediction and Detection

Several different machine learning techniques or algorithms or models, some of which have been already discussed earlier in the chapter are being actively used by the researchers for the detection and prediction of different diseases. A review and comparison of these techniques is done in Table 1.

5 DL Models Used for Disease Prediction and Detection

As discussed earlier, DL is a subset of ML and is very useful, when it comes to task of computer vision or image processing. Hence, many researchers have shown promising results for prediction and diagnosis of different diseases such as diabetes, (as depicted by Luis Fregoso-Aparicio et al. [17] and Saritha Balasubramaniyan [18]) and Brain tumors (as depicted in studies from [19–23]). Different DL architectures such as DNN, CNN, ANN and others are commonly used for predicting different diseases, some of which are compared in the Table 2.

6 New Emerging Techniques for Medical Usage

Nowadays, more advanced and emerging technologies including the Virtual reality (VR), Augmented reality (AR), and Mixed Reality (MR) are being developed specifically aligned with the medical use cases, which includes the training of students, AI assisted surgeries and disease detection and prediction. These technologies have not yet been fully adopted in the hospitals and is in very early stage of the development and research [43].

Table 1 Review of different ML models being used for diagnosis of different diseases

Reference number	Author name	Technique used	Conclusion
[15]	Mustafa Abdul Salaam et al.	Federated Machine Learning and Traditional Machine Learning	Using Federated ML has higher prediction and accuracy Federated ML has less loss than the traditional model Federated ML takes more time to train the model than the traditional model
[16]	Luca Brunese et al.	Supervised Machine Learning	Effectiveness of the proposed method by obtainingan average precision value and recall value is equal to 0.965 for the discrimination of COVID-19 and other pulmonary diseases having same or similar symptoms
[17]	Sun L et al.	Support vector Machine	Accuracy: 77.5%
[18]	Wu J et al.	Random Forest	Accuracy: 95.95%
[19]	Yan L et al.	XGBoost Classifier	Accuracy: 90%
[20]	Loey Mohamed et al.	ResNet 18	Accuracy: 100%
[21]	ErdiAcar et al.	DenseNet 169	Accuracy: 99.80%.highl y accurate and precise method
[22]	Chowdhury MEH et al.	DenseNet 201	Accuracy: 99.70%
[23]	Toğaçar M et al.	SqueezeNet and MobileNetV2	Accuracy: 99.27%
[24]	Song Y et al.	ResNet50	Accuracy: 86%
[21]	Erdi Acar et al.	VGG 16	Accuracy: 99.20%
[25]	Ardakani A et al.	AlexNet	Accuracy: Training: 82.60%; Validation: 78.92%
[21]	Erdi Acar et al.	DenseNet 121	Accuracy: 99.45%
[26]	Kassani SH et al.	Inception V3	Accuracy: 95%
[27]	Zhang HT et al.	MADE-DBM	Accuracy: 96.19%
[28]	Hu S et al.	Weekly Supervised DL	Accuracy: 96%
[29]	Ashwini R et al.	Support Vector Machine and Logistic Regression	Accuracy: 98%
[30]	Abid Sarwar et al.	Support Vector Machine and Naive Bayes	Accuracy: 97.34%

Table 2 Review of different DL models being used for prediction of different diseases

Reference number	Author	Technique used	Conclusion
[31]	Luis Fregoso-Aparicio et al.	Deep Neural Network	Better results are found by using various techniques
[32]	Saritha Balasubramaniyan	Neural Network	Extracted features for better classification accuracy and reduced result rate
[33]	DC Febrianto et al.	Convolutional Neural Network	Accuracy: 0.93
[34]	A. Yang et al.	Convolutional Neural Network	Accuracy: 0.99
[35]	M. Sharma et al.	Discrete Wavelet Transformation + Principal Component Analysis + K-Nearest Neighbor	Accuracy: 0.97
[13]	S. Alsubai et al.	Convolutional Neural Network + Long Short Term Memory	Accuracy: 0.99
[36]	D. Mukherkjee et al.	Regions with Convolutional Neural Network and Support Vector Machine	Accuracy: 0.95
[37]	A. Reiling, et al.	Convolutional Neural Network + Genetic Algorithm	Accuracy: 0.85
[38]	P.A. Tjahyaningtijas H et al.	Encoded Convolutional Neural Network	Accuracy: 0.97
[39]	H. H. Sultan et al.	Deep Neural Network	Accuracy: 0.96
[40]	P. K. Mallick et al.	Deep Wavelet Autoencoder based Deep Neural Network	Accuracy: 0.96
[41]	P. M. Shakeel et al.	Machine Learning Based Back Propagation Neural Network	Accuracy: 0.98
[42]	P Gokila Brindha et al.	Artificial Neural Network	Accuracy: 0.65

6.1 Augmented Reality

AR is a new age technology, which nowadays has become a hot topic of discussion among public. It is a technology, which expands the user's reality using the virtual or digital information [44]. Making it simpler, we can say in AR, users work alongside with the digital objects that are kept in the real world [45]. In medical field, AR technology can be used for assisting and planning the surgeries along with the demonstration and teaching of anatomical structures to the medicine students, which can help them understand the internal body organs more clearly [43–45]. In teaching, using this technique has many advantages, which includes:

1. AR provides three dimensional models, which almost replicates an original organ in virtual environment, that covers up the teaching- understanding gaps and therefore creating better atmosphere for understanding the concepts in details.
2. The other major benefit of this technique is the interactivity of the tool with the user, with which, students can focus on any of the given portion, which is being displayed on the screen. Moreover, students can rotate the screen and also zoom in and zoom out as per convenience making it even more easier in understanding.
3. Anytime and anywhere access makes it even better than using the old age technologies such as textbooks and learning through presentations.

6.2 Virtual Reality

It is a type of simulating technology, in which the user can interact and move with virtual space, events and objects in a virtual or digital environment. The user in VR feels as if he is totally immersed in the environment, giving him more realistic feel. In medical domain, VR is being used in various ways such as for the motor rehabilitation, therapies and educational use cases [46]. Table 3 brings in a summary literature for AR and VR technologies being used in the medical field including procedures.

Table 3 Summary of literature review and outcomes of using AR and VR technologies

Reference number	Author	Procedure	Technique used	Conclusion
[47]	Elmi-Terander et al.	Transpedicular screw	VR simulation and AR-guide	Good accuracy, reduced time and good clinical outcomes
[48]	Edström et al.	Transpedicular screw	VR simulation and AR-guide	Ratiation exposure, clinical outcomes
[49]	Carl et al.	Spinal tumor resection (whole spine)	AR-guide	Feasibility of intraoperative use
[50]	Umebayashi et al.	Cervical foraminotomy	AR-guide	Feasibility of intraoperative use
[51]	Farber et al.	Lumber vertebral body puncture	VR-simulation	Good accuracy

7 Security and Privacy Risks and Challenges

Security and Privacy risks prevail as we try to incorporate the use of AI in healthcare. Using AI on one hand makes things easier and faster but is also very volatile to the security challenges including the data prevailing to the patient, hence making people more resilient to using the AI in healthcare. Some of the security challenges occur while implementing the which are discussed below.

1. **Transparency of patient data**: One of the biggest challenges faced includes is to maintain the transparency of the data collected of the patient by the AI models to perform the diagnosis and other tests. This is one of the basic rights of the patient to know how and which data of the patient is being used. Maintaining this transparency is very difficult using the AI models.
2. **Data migration**: Other major challenge that is faced is the migration of currently available data to cloud and other systems that are essential for AI models to work. This process will require a lot of manpower and is not an easy task as currently most of the data available is in the form of hard texts, which needs to be converted to softcopies and some of the formatting might also be required to change in order to successfully migrate the existing data.
3. **Upskilling the current workforce to make them familiar with AI**: The current workforce including the physicians, technicians and other support staff needs to be trained in how to use the AI based systems. This training would require a technical understanding of the AI system alongside familiarity with the technical knowledge of the domain.
4. **Ransomware and malware based cyber-attacks**: AI models are very vulnerable to different kinds of cyber-attacks. These attacks could be of any type, but the most common types include AI phishing, Evasion, poisoning, privacy, and abuse. AI phishing is a type of attack, in which the users are made to believe that the signals, messages and other outputs visible are coming from the verified or authorized source, but in actual they are not from any legitimate source, this kind of attacks cause the data breach and digital spying.
5. **Problem of reidentification**: This also relates to the external risk of privacy breaches. Breach of data related to healthcare has seen a rise around the world, although not by the cyber criminals currently [52], but could be used by such people soon. Although the data used by AI models is most of the time anonymous and does not contain any personal information of the patients, but studies shows that the identification can be regenerated by using algorithms [52]. For instance, study conducted by Liangyuan Na et al. [53] concluded that using Random Forest algorithm, 94.9% of data of adults was being reidentified whereas 87.4% of the children's data could be reidentified. Even some of the studies have successfully showcased how an algorithm can be used to reidentify unknown health data which would link the data available on internet to that with the people in the existence [52].

Discussed above are some of the challenges and risks of using AI models particularly in healthcare sector, due to which such systems have not yet been fully adopted and people resist using such systems in real life. One other most contributing factor include the conservative thinking of people, due to which, the common public is not ready to believe that computers and AI have the capability to properly detect and, in many cases, have proved to even treat the problem and cure the patient.

8 Conclusion

This chapter highlights the transformative role of AI and ML in enhancing diagnostic accuracy and efficiency in healthcare, especially in disease prediction and early detection. With advancements in ML and DL, healthcare has witnessed significant improvements in the analysis of medical imaging, contributing to quicker and more accurate diagnoses. These technologies, ranging from supervised ML algorithms like Random Forest and Support Vector Machines to deep learning models such as CNN and LSTM, showcase promising potential in tackling complex medical challenges, particularly in cancer and other critical disease detection. Despite the successes, there remain significant challenges in integrating AI into healthcare. Security and privacy risks, the need for data transparency, and the imperative of upskilling healthcare professionals present barriers to widespread adoption. Furthermore, the evolving threats of cyber-attacks and data breaches require robust protective measures to ensure patient information remains confidential. Emerging technologies like augmented and virtual reality provide exciting opportunities for education and procedural support, further bridging the gap between theoretical knowledge and practical application. As AI continues to evolve, ongoing research and development will be essential in overcoming these challenges, promoting trust in AI systems, and refining their applications for broader, safer, and more effective use in healthcare.

References

1. Kumar, N., Narayan Das, N., Gupta, D., Gupta, K., Bindra, J.: Efficient automated disease diagnosis using machine learning models. J. Healthc. Eng. **2021**, 1–13 (2021). https://doi.org/10.1155/2021/9983652
2. Kaur, D., Ummat, A., Kaur, S.: Comparative study of machine learning based tumor detection techniques in human brain. In: 2021 2nd International Conference on Computational Methods in Science & Technology (ICCMST), pp. 118–122. IEEE (2021). https://doi.org/10.1109/ICCMST54943.2021.00034
3. Kaur, S., Kumar, Y., Koul, A., Kumar Kamboj, S.: A systematic review on metaheuristic optimization techniques for feature selections in disease diagnosis: open issues and challenges. Arch. Comput. Methods Eng. **30**(3), 1863–1895 (2023). https://doi.org/10.1007/s11831-022-09853-1
4. Bedi, H.P.S., Kaur, S., Rajvanshi, S.: Tumor detection and classification. In: Bio-Inspired Optimization for Medical Data Mining, pp. 75–102. Wiley (2024). https://doi.org/10.1002/9781394214211.ch4

5. Uddin, S., Khan, A., Hossain, M.E., Moni, M.A.: Comparing different supervised machine learning algorithms for disease prediction. BMC Med. Inform. Decis. Mak. **19**(1), 281 (2019). https://doi.org/10.1186/s12911-019-1004-8
6. Ali, M.M., Paul, B.K., Ahmed, K., Bui, F.M., Quinn, J.M.W., Moni, M.A.: Heart disease prediction using supervised machine learning algorithms: performance analysis and comparison. Comput. Biol. Med. **136**, 104672 (2021). https://doi.org/10.1016/j.compbiomed.2021.104672
7. Gupta, C., Saha, A., Subba Reddy, N.V., Dinesh Acharya, U.: Cardiac disease prediction using supervised machine learning techniques. J. Phys. Conf. Ser. **2161**(1), 012013 (2022). https://doi.org/10.1088/1742-6596/2161/1/012013
8. Sazzadur Rahman, A.K.M., Javed Mehedi Shamrat, F.M., Tasnim, Z., Roy, J., Hossain, S.A.: A comparative study on liver disease prediction using supervised machine learning algorithms. Int. J. Sci. Technol. Res. **8**(11), 419–422 (2019)
9. Barlow, H.B.: Unsupervised learning. Neural Comput. **1**(3), 295–311 (1989). https://doi.org/10.1162/neco.1989.1.3.295
10. Dirid, S.: Unsupervised learning—A systematic literature review. OSF Preprints (2021)
11. Alzubaidi, L., et al.: Review of deep learning: concepts, CNN architectures, challenges, applications, future directions. J Big Data **8**(1), 53 (2021). https://doi.org/10.1186/s40537-021-00444-8
12. Lotlikar, V.S., Satpute, N., Gupta, A.: Brain tumor detection using machine learning and deep learning: a review. Curr. Med. Imaging Formerly Curr. Med. Imaging Rev. **18**(6), 604–622 (2022). https://doi.org/10.2174/1573405617666210923144739
13. Alsubai, S., Khan, H.U., Alqahtani, A., Sha, M., Abbas, S., Mohammad, U.G.: Ensemble deep learning for brain tumor detection. Front Comput Neurosci **16** (2022). https://doi.org/10.3389/fncom.2022.1005617
14. Shanthi, S., Saradha, S., Smitha, J.A., Prasath, N., Anandakumar, H.: An efficient automatic brain tumor classification using optimized hybrid deep neural network. Int. J. Intell. Netw. **3**, 188–196 (2022). https://doi.org/10.1016/j.ijin.2022.11.003
15. Abdul Salam, M., Taha, S., Ramadan, M.: COVID-19 detection using federated machine learning. PLoS One **16**(6), e0252573 (2021). https://doi.org/10.1371/journal.pone.0252573
16. Brunese, L., Martinelli, F., Mercaldo, F., Santone, A.: Machine learning for coronavirus covid-19 detection from chest X-rays. Procedia Comput. Sci. **176**, 2212–2221 (2020). https://doi.org/10.1016/j.procs.2020.09.258
17. Sun, L., et al.: Combination of four clinical indicators predicts the severe/critical symptom of patients infected COVID-19. J. Clin. Virol. **128**, 104431 (2020). https://doi.org/10.1016/j.jcv.2020.104431
18. Jiangpeng, et al., W.: Rapid and accurate identification of COVID-19 infection through machine learning based on clinical available blood test results. medRxiv preprint (2020). https://doi.org/10.1101/2020.04.02.20051136
19. Yan, L., et al.: An interpretable mortality prediction model for COVID-19 patients. Nat. Mach. Intell. **2**(5), 283–288 (2020). https://doi.org/10.1038/s42256-020-0180-7
20. Loey, M., Smarandache, F., Khalifa, N.E.M.: Within the lack of chest COVID-19 X-ray dataset: a novel detection model based on GAN and deep transfer learning. Symmetry (Basel) **12**(4), 651 (2020). https://doi.org/10.3390/sym12040651
21. Acar, E., Şahin, E., Yılmaz, İ: Improving effectiveness of different deep learning-based models for detecting COVID-19 from computed tomography (CT) images. Neural Comput. Appl. **33**(24), 17589–17609 (2021). https://doi.org/10.1007/s00521-021-06344-5
22. Chowdhury, M.E.H., et al.: Can AI help in screening viral and COVID-19 pneumonia? IEEE Access **8**, 132665–132676 (2020). https://doi.org/10.1109/ACCESS.2020.3010287
23. Toğaçar, M., Ergen, B., Cömert, Z.: COVID-19 detection using deep learning models to exploit Social Mimic Optimization and structured chest X-ray images using fuzzy color and stacking approaches. Comput. Biol. Med. **121**, 103805 (2020). https://doi.org/10.1016/j.compbiomed.2020.103805

24. Song, Y., et al.: Deep learning enables accurate diagnosis of novel coronavirus (COVID-19) with CT images. IEEE/ACM Trans. Comput. Biol. Bioinform. **18**(6), 2775–2780 (2021). https://doi.org/10.1109/TCBB.2021.3065361
25. Ardakani, A.A., Kanafi, A.R., Acharya, U.R., Khadem, N., Mohammadi, A.: Application of deep learning technique to manage COVID-19 in routine clinical practice using CT images: results of 10 convolutional neural networks. Comput. Biol. Med. **121**, 103795 (2020). https://doi.org/10.1016/j.compbiomed.2020.103795
26. Kassani, S.H., Kassasni, P.H., Wesolowski, M.J., Schneider, K.A., Deters, R.:Automatic Detection of coronavirus disease (COVID-19) in X-ray and CT Images: a machine learning-based approach (2020)
27. Zhang, H., et al.: Automated detection and quantification of COVID-19 pneumonia: CT imaging analysis by a deep learning-based software. Eur. J. Nucl. Med. Mol. Imaging **47**(11), 2525–2532 (2020). https://doi.org/10.1007/s00259-020-04953-1
28. Hu, S., et al.: Weakly supervised deep learning for COVID-19 infection detection and classification from CT images. IEEE Access **8**, 118869–118883 (2020). https://doi.org/10.1109/ACCESS.2020.3005510
29. Almahdawi, A., Naama, Z.S., Afshin, K.V., Prof. Raj, D.: Diabetes prediction using machine learning. Int. J. Res. Appl. Sci. Eng. Technol. **10**(4), 544–549 (2022). https://doi.org/10.22214/ijraset.2022.41143
30. Sarwar, A., Ali, M., Manhas, J., Sharma, V.: Diagnosis of diabetes type-II using hybrid machine learning based ensemble model. Int. J. Inf. Technol. **12**(2), 419–428 (2020). https://doi.org/10.1007/s41870-018-0270-5
31. Fregoso-Aparicio, L., Noguez, J., Montesinos, L., García-García, J.A.: Machine learning and deep learning predictive models for type 2 diabetes: a systematic review. Diabetol. Metab. Syndr. **13**(1), 148 (2021). https://doi.org/10.1186/s13098-021-00767-9
32. Balasubramaniyan, S., Jeyakumar, V., Nachimuthu, D.S.: Panoramic tongue imaging and deep convolutional machine learning model for diabetes diagnosis in humans. Sci. Rep. **12**(1), 186 (2022). https://doi.org/10.1038/s41598-021-03879-4
33. Febrianto, D.C., Soesanti, I., Nugroho, H.A.: Convolutional neural network for brain tumor detection. IOP Conf Ser Mater Sci Eng **771**(1), 012031 (2020). https://doi.org/10.1088/1757-899X/771/1/012031
34. Yang, A., Yang, X., Wu, W., Liu, H., Zhuansun, Y.: Research on feature extraction of tumor image based on convolutional neural network. IEEE Access **7**, 24204–24213 (2019). https://doi.org/10.1109/ACCESS.2019.2897131
35. Sharma, M., Dr. Mukharjee, S.: Artificial neural network fuzzy inference system (ANFIS) for brain tumor detection. ArXiv preprint (2012)
36. Mukherkjee, D., Saha, P., Kaplun, D., Sinitca, A., Sarkar, R.: Brain tumor image generation using an aggregation of GAN models with style transfer. Sci. Rep. **12**(1), 9141 (2022). https://doi.org/10.1038/s41598-022-12646-y
37. Reiling, A., Mitchell, W., Westberg, S., Balster, E., Taha, T.: CNN optimization with a genetic algorithm. In: 2019 IEEE National Aerospace and Electronics Conference (NAECON), pp. 340–344. IEEE (2019). https://doi.org/10.1109/NAECON46414.2019.9058307
38. Purnomo, M.: Brain tumor classification in MRI images using En-CNN. Int. J. Intell. Eng. Syst. **14**(4), 437–451 (2021). https://doi.org/10.22266/ijies2021.0831.38
39. Sultan, H.H., Salem, N.M., Al-Atabany, W.: Multi-classification of brain tumor images using deep neural network. IEEE Access **7**, 69215–69225 (2019). https://doi.org/10.1109/ACCESS.2019.2919122
40. Kumar Mallick, P., Ryu, S.H., Satapathy, S.K., Mishra, S., Nguyen, G.N., Tiwari, P.: Brain MRI image classification for cancer detection using deep wavelet autoencoder-based deep neural network. IEEE Access **7**, 46278–46287 (2019). https://doi.org/10.1109/ACCESS.2019.2902252
41. Mohamed Shakeel, P., El Tobely, T.E., Al-Feel, H., Manogaran, G., Baskar, S.: Neural network based brain tumor detection using wireless infrared imaging sensor. IEEE Access **7**, 5577–5588 (2019). https://doi.org/10.1109/ACCESS.2018.2883957

42. Gokila Brindha, P., Kavinraj, M., Manivasakam, P., Prasanth, P.: Brain tumor detection from MRI images using deep learning techniques. IOP Conf. Ser. Mater. Sci. Eng. **1055**(1), 012115 (2021). https://doi.org/10.1088/1757-899X/1055/1/012115
43. Bedi, H.P.S.: Image processing—A tool for medical diagnostics. Geeks For Geeks
44. Eckert, M., Volmerg, J.S., Friedrich, C.M.: Augmented reality in medicine: systematic and bibliographic review. JMIR Mhealth Uhealth **7**(4), e10967 (2019). https://doi.org/10.2196/10967
45. Sethiya, K., Guruprasad, M.: Augmented reality (AR) in healthcare. J Interdiscipl. Cycle Res. **XII**, 343–359 (2021)
46. Kamińska, D., Zwolińsksi, G., Laska-Leśniewicz, A., Coelho, L.P.: Virtual reality in healthcare, pp. 1–10 (2022). https://doi.org/10.4018/978-1-7998-8371-5.ch001
47. Elmi-Terander, et al., A.: Pedicle screw placement using augmented reality surgical navigation with intraoperative 3D imaging. Spine (Phila Pa 1976), **44**(7), 517–525 (2019). https://doi.org/10.1097/BRS.0000000000002876
48. Edström et al., E.: Augmented reality surgical navigation in spine surgery to minimize staff radiation exposure. Spine (Phila Pa 1976) **45**(1), E45–E53 (2020). https://doi.org/10.1097/BRS.0000000000003197
49. Carl, B., Bopp, M., Saß, B., Voellger, B., Nimsky, C.: Implementation of augmented reality support in spine surgery. Eur. Spine J. **28**(7), 1697–1711 (2019). https://doi.org/10.1007/s00586-019-05969-4
50. Umebayashi, D., Yamamoto, Y., Nakajima, Y., Fukaya, N., Hara, M.: Augmented reality visualization–guided microscopic spine surgery: transvertebral anterior cervical foraminotomy and posterior foraminotomy. JAAOS: Glob. Res. Rev. **2**(4), e008 (2018). https://doi.org/10.5435/JAAOSGlobal-D-17-00008
51. Färber, M., Hummel, F., Gerloff, C., Handels, H.: Virtual reality simulator for the training of lumbar punctures. Methods Inf. Med. **48**(05), 493–501 (2009). https://doi.org/10.3414/ME0566
52. Murdoch, B.: Privacy and artificial intelligence: challenges for protecting health information in a new era. BMC Med. Ethics **22**(1), 122 (2021). https://doi.org/10.1186/s12910-021-00687-3
53. Na, L., Yang, C., Lo, C.-C., Zhao, F., Fukuoka, Y., Aswani, A.: Feasibility of reidentifying individuals in large national physical activity data sets from which protected health information has been removed with use of machine learning. JAMA Netw. Open **1**(8), e186040 (2018). https://doi.org/10.1001/jamanetworkopen.2018.6040

Hermehar P. S. Bedi is a Ph.D. Scholar at IIIT Delhi, India, specializing in Extended Reality. His research interests include artificial intelligence in health technology and medical image processing. With seven publications, including papers, book chapters, and articles in esteemed national and international journals and conferences, his work is featured in prominent outlets such as Wiley, Springer, and Taylor & Francis.

Renu Puri earned her Ph.D. from IKG Punjab Technical University, Jalandhar, and brings over 19 years of extensive experience as an academician and researcher. Currently an Associate Professor at an engineering college, she also serves as the Research Coordinator for M.Tech theses. Dr. Puri holds multiple professional memberships and has published over 40 papers in various Scopus-indexed international and national journals, as well as IEEE conferences. She has filed two patents, with five more under review. Her research focuses on soft computing, swarm intelligence, and antenna optimization.

Amandeep Kaur is an Assistant Professor in the Department of Computer Science and Engineering at Chandigarh Engineering College, Landran, Mohali, India. She holds a Master's degree in Engineering in Information Technology, with a specialization in Digital Image Processing, completed in 2014. Her research expertise lies in Image Processing, with interests extending to Machine Learning and Artificial Intelligence. Ms. Kaur has authored over 10 research papers in various international journals and published a book titled Compression Technique for Land Data Recording with Lambert Publications.

Deepak Puri is an accomplished academic and researcher with over 12 years of experience spanning industry and academia. Currently an Associate Professor in Pharmaceutics at a Pharmacy College, he has a strong record of publications in both national and international journals, as well as an Indian patent to his credit. His research interests focus on characterizing the physical properties of drugs and developing innovative drug delivery systems, contributing significantly to advancements in pharmaceutical sciences.

Comprehensive Analysis of Machine Learning Approaches for Identifying Polycystic Ovary Syndrome

Mehak Abbas and Shaheen Naz

Abstract This chapter examines the rising incidence of Polycystic Ovary Syndrome (PCOS), a prevalent hormonal disorder with significant implications for women's health, particularly in relation to infertility. Early detection of PCOS is essential for effective management and minimizing complications. As machine learning (ML) has demonstrated superior accuracy in feature extraction for diagnostic purposes, various ML models, including Convolutional Neural Networks (CNN), Support Vector Machines (SVM), K-Nearest Neighbors (KNN), Random Forest (RF), Logistic Regression (LR), Decision Trees (DT), and Naive Bayes (NB), are explored for PCOS detection. This chapter provides an in-depth analysis of ML-based techniques for identifying PCOS, detailing the effectiveness of each approach based on extensive research from recent decades. It reviews datasets commonly used in PCOS studies, offering quantitative and qualitative performance comparisons among algorithms. The chapter concludes by highlighting key challenges in current methodologies and identifying promising directions for future research in ML-based PCOS detection.

Keywords PCOS detection · Machine learning · Early diagnosis · Women's health · Infertility · Convolutional neural networks · Support vector machines · Random forest · Healthcare technology · Future research

1 Introduction

The female reproductive system is prone to several diseases that might eventually result in major health problems. Ovaries, uterus, cervix, vagina, etc. are all involved in these medical conditions. Hormonal imbalances, abnormal sleep habits, stress, and other factors are the main causes of these disorders' occurrence. Under the classification of hyperandrogenism, or excessive androgen production by the ovaries,

M. Abbas (✉) · S. Naz
Department of Electrical and Electronics Communication Engineering, Sharda University, Greater Noida, India
e-mail: mehakabbas762@gmail.com

N. R. Vajjhala et al. (eds.), *Artificial Intelligence in Healthcare Information Systems—Security and Privacy Challenges*, Information Systems Engineering and Management 34, https://doi.org/10.1007/978-3-031-84404-1_3

is polycystic ovarian syndrome, or PCOS. The female reproductive age group (ages 15 to 40) is most susceptible to this condition. This age group has been identified when women enter menopause after the age of 40, and menstruation starts before the age of 15. This means that irregular menstrual cycles are quite likely. Women's hormone levels are impacted by this disorder [1]. Ovaries are classified as normal, cystic, or polycystic depending on their structural characteristics [2]. Every month, cysts carrying eggs form in normal ovaries [3]. Every ovarian cyst that develops has water inside of it, which is released from the ovary during menstruation. The remaining cysts in ovarian tissue give rise to polycystic ovaries when they are unable to be excluded from the body. In ovaries with polycystic ovarian syndrome, ovulation is not possible despite the large number of follicles present.

Between polycystic and normal ovaries, this is the primary distinction [4]. It affects around 20% of women who can procreate [5]. It is frequently correlated with increased body levels of two male hormones. This hormone makes it more difficult for them to become pregnant and causes their bodies to skip menstruation cycles. Women suffering from PCOS create more male hormones than normal. The PCOS side effects manifest in a variety of unique ways. While some women will only have a few moderate or mild adverse effects, other women will experience several major negative effects. Some of the adverse symptoms of PCOS include mood swings, depression, unease, and poor self-esteem. Additionally, signs might change throughout various times of a woman's life. PCOS's main signs include paused times, undeveloped ovulation eggs which don't ovulate, various ovarian "cysts" difficulty becoming a mother, obesity, hair loss, generally on the sides of the lips, the body, lower abdomen, or bottoms, abnormal growth of hair, acne, greasy skin as well as undesirable locks through the top of the head these are a few of the symptoms [6]. More signs of PCOS include being overweight, skin discoloration and pigmentation of the skin. The consequences of PCOS include diabetic complications, cardiovascular illness, breathing difficulties during sleeping, fatigue, and early abortion. The following diagram makes it simple for researchers to identify the signs and consequences of PCOS. Figures 1 and 2 illustrate the signs and consequences of PCOS.

Unfortunately, PCOS is very challenging for doctors to correctly diagnose at an early stage because of its vast spectrum of symptoms and the presence of other concurrent gynecological issues [7]. Even though the condition's specific etiology is unknown some researchers think that the primary issue is excessive androgen synthesis in ovarian theca cells [8, 9]. The diagnostic criteria for this hormonal condition are highly disputed, however, the norms that were applied during the 2003 session in Rotterdam are typically regarded as the clinical confirmation of PCOS [10]. Two of the three requirements stated below can be used to identify PCOS, based on Rotterdam Diagnostic Factors: A clinical or biochemical sign that elevated androgen activity, oligo/amenorrhea and PCO follicles are the three main conditions as determined by the results of Ultrasonography (USG). However, human vision is still used for the manual evaluation of USG data at present [11, 12]. The most significant thing is to identify between people with PCOS and those without and to start treating Polycystic Ovary Syndrome as soon as feasible. Given that the situation resulted in an ovarian disease, which raises the risk of pregnancy troubles and mental

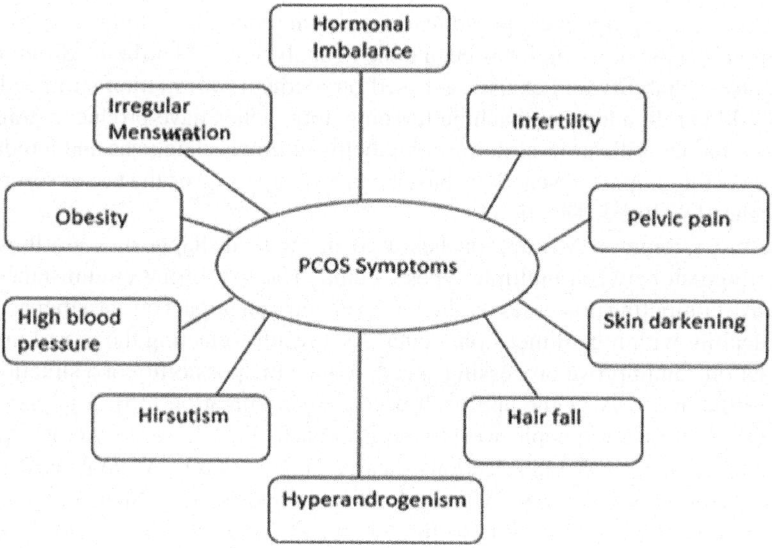

Fig. 1 Symptoms that are present in PCOS-affected women

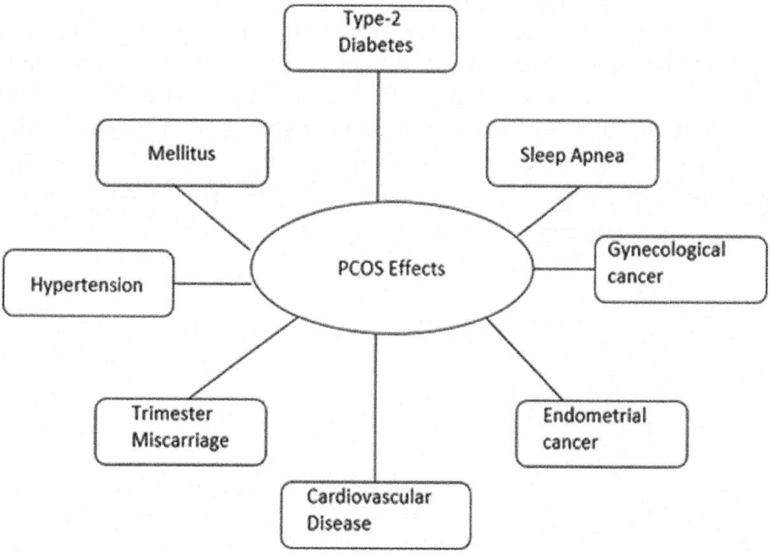

Fig. 2 Some fundamental effects of PCOS in women

health issues, it requires every possible test including imaging techniques as possible. Even though a lot of research has been done to analyze PCOS utilizing different ML techniques, reliability and exactness based on medical information still need to be improved [13]. In addition, machine learning approaches have produced gratifying results in the identification of disease that helps in timely diagnosis and a reduction in the mortality impact, researchers have been moving towards the use of several ML approaches for illness detection [14].

The most common ML approaches used in medical diagnostics are those that can distinguish between multiple types of data. The overfitting circumstances and expensive computing processes might, however, have an impact on classifier accuracy when dealing with high dimensional data. As a result, selecting the most important information can improve processing speed, reduce the possibility of a situation, and enhance the accuracy of the approach used for classification [15]. The prevalence of PCOS as it affects people worldwide is seen in Fig. 3. Asian women had the highest prevalence of PCOS infection—nearly 31.3%—while White Americans had a lower rate of 4.8%. 6.8% of African Americans and 6.8% of Spanish women are afflicted respectively. According to this graph, PCOS in women is mostly caused by two hereditary and racial factors. The authors considered every research that has been published between 2015 and 2022 that specifically addressed applying machine learning to identify and forecast PCOS.

The Evaluations of publications mentioned above that utilize statistical parameters for PCOS identification using ML do not exist, though. The authors are aware that there are only a limited number of research papers about Polycystic Ovary Syndrome identification. Some employ machine learning (ML) directly, while others

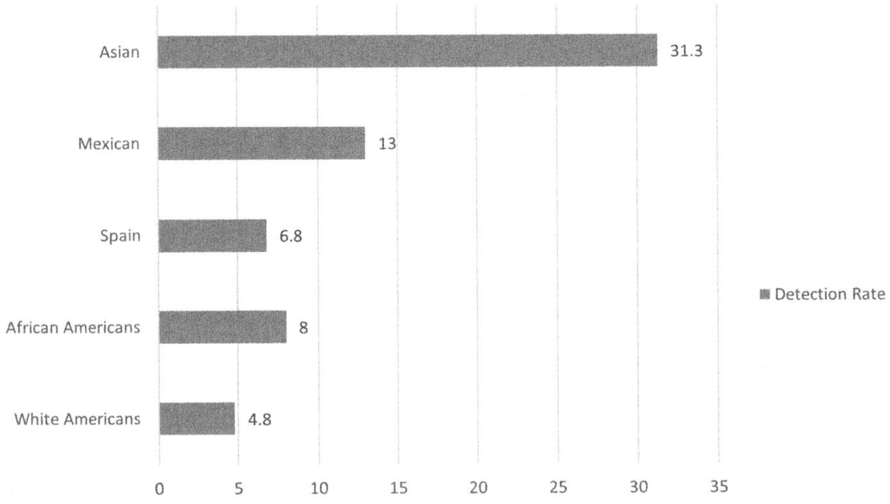

Fig. 3 Global variation in polycystic ovary syndrome Detection Rates across ethnically diverse women

use computer-assisted technologies, image processing, and artificial intelligence (AI).

Gotte and Jyothi [16] conducted research on PCOS detection segmentation and classifications using several machine learning (ML) methods. However, there is no mention of a prospective literature comparison in the survey. Papers that were released between 2020 and 2021 were examined by Vaswani et al. [17]. The study offers an overview of the studies that identify PCOS using ML methods. They took into account the studies that employ ML and DL for Polycystic Ovary Syndrome detection. These individuals were the ones who thought of employing the identification of things to find PCOS. Research studies that are linked to image processing were examined by Pulluparambil and Bhat [18]. They performed a detailed analysis of the detecting methods, stages, and algorithms currently in use. They identified several research gaps and suggested a PCOS detection framework. They performed a detailed analysis of the detecting methods, stages, and algorithms currently in use. They identified several research holes and put up a PCOS detection framework. The research publications that employed computer-aided methods for follicle recognition and PCOS diagnosis in ovarian ultrasound pictures were examined by Rabiu et al. [19]. They looked at articles published between 1997 and 2014. They analyzed the methods already in use to wrap up their research. They divided the techniques into five pieces and talked about the procedure. Agrawal et al. [20] updated AI detection algorithms for PCOS, displaying the results without comparing them. The current research papers suffer from significant deficiencies. In the beginning, they simply assessed a small number of articles and focused on those studies, neglecting to address any systematic data, such as flaws and performance comparisons. Secondly, they didn't demonstrate either numerical or artistic evaluation of the procedures. Lastly, researchers failed to offer a complete set of data analysis or a sufficient plan for the scope of future studies. This paper explores the identification of Polycystic Ovary Syndrome and provides an analytical approach to understanding the algorithms used in modern techniques. It helps researchers select the optimal algorithm for their application and provide a precise explanation of the relevant collection of data. Using all of the previous studies on Polycystic Ovary Syndrome identification techniques, the paper also highlights the variety and unresolved issues in the field.

2 Evaluation of Published Works

The recently published research, employed approaches, and algorithms are all covered in the section that follows. To provide a quick assessment of the research, an informative comparison, and a table are also included. In addition, there is a table in this part of the paper that lists all the works that are currently in existence and provides suggestions on each work. According to Sumathi et al. [21] Feature extraction for PCOS using CNNs has been suggested to analyze pictures in which a cyst can be detected using an algorithm that depends on the extraction of image processing features. OpenCV is used for measuring the variables while the Watershed technique is used for extracting features. The system is limited to identifying cysts; it is unable to detect the presence or absence of PCOS. 85% accuracy is displayed by the model. In a 5-overlap cross-validation, CNN with a f1-rating of 100% and an average rate of 76.36%. A range of ML methods, including RF, SVM CART, Naive Bayes (NB), and LR, were used by Hassan and Mirza [22] to categorize the Kaggle PCOS datasets. Moreover, testing data showed that RF might beat other classifiers by obtaining a 96% classification accuracy rate. A technique for automated feature derivation utilizing a Convolutional Neural Network with an f1-Score (100%) and a regularity (76.36%) on 5-overlapping cross-validation was proposed by Cahyono et al. [23]. There are 40 Non-Polycystic Ovary data overall in the collection, along with 14 data related to polycystic ovarian disease (PCO). Islam et al. [24], the goal was to identify PCOS using ovarian ultrasonography. They used a CNN (VGG-16) and ensemble models (XGBoost) and achieved an impressive 99.89% accuracy. The DNN method for feature extraction outperformed others in accuracy, precision, recall, and F1 score, highlighting the effectiveness of stacking ensemble machine learning for PCOS detection.

To identify females who have PCOS. Mehr and Polat [25] used a duo, conventional, and applied algorithms. A set of data containing all characteristics and limited portions of attributes, It was developed with the help of filtering, insertion, and covering feature selection approaches. The method was used to investigate multiple classification techniques, such as MLP, Ensemble RF, Extra Tree, and AdaBoost. Gupta et al. [26] offered a discriminant study in various measurements using exponential and quadratic forms for the classification of binary data. Using quadratic discriminant analysis, training accuracy achieved 97.37%, While evaluation, the precision achieved 95.92%. The analysis area still needs improvement. The paper discusses an enhanced collection machine learning (ML) classifier strategy for detecting PCOS from people's illness information with 5 conventional Machine Learning approaches that serve as initial newcomers, with one bagged or boosted aggregate model of ML acting as a meta-learner. Vasavi et al. [27] created a classification-based system technique to identify PCOS. Finally, they defined a hardware requirement and created an app specifically for PCOS patients. They applied Decision Tree, K-nearest neighbor, Logistic Regression, Naïve Bayes, Support Vector Machine, and Random Forest for PCOS detection. The feature was chosen after the data had been preprocessed before the methods were used. Deshpande and Wakankar [28] suggested a computerized approach for identifying PCOS by counting the number of follicles in an image of the ovary and then combining clinical, biochemical, and imaging characteristics to divide patients into two groups, i.e., those without PCOS and those with it. The model comprises selecting attributes, preprocessing, the technique of segmentation, and prediction. In categorization, the Support Vector Machine approach is employed. Khanna et al. [29] examined a technique that involved evaluating and examining 12 different ML techniques including LR, DT, RF, SVM, NB, KNN, Adaptive Boosting, gradient Boosting and Extra trees. It produced three stacks. Seven classifiers were mixed using STACK-1, including the Logistic Regression, Support Vector Machine (having units that are linear, polynomial, gaussian, and spherical), Naïve Bayes, and K-Nearest Neighbor techniques. A group of tree-based classifiers, including DT, RF, Adaptive Boosting, gradient Boosting and Extra trees, were represented by STACK-2. STACK- 1 and STACK-2 were stacked on top of one another to create STACK-3. The last multi-stack of machine learning algorithms worked most effectively, having precision, reliability, recall, and F1-scores of 98%, 97%, 98%, and 98%, respectively. To automate PCOS identification, Dutta et al. [30] intended to use the Synthetic Minority Oversampling Technique and 5 Machine learning approaches, including Logistic Regression, Decision Tree, Support Vector Machine and K-Nearest Neighbor. The suggested LR outperformed previous algorithms at accurately identifying datasets.

The Nave Bayes (NB) classifier and ANN were demonstrated by Maheshwari et al. [31] for PCOS diagnosis. In addition, we may use this to identify the motivating factors and undesirable from PCOS in previous setups depending on the data collected. Efficiency is around 98.63%, reliability and sensitivity are 100%, the F-value is 68.76%, and the overall rating is 55%. To identify PCO from ovarian ultrasound images, Prapty et al. [32] developed the CNN-based model PCONet. InceptionV3, a 45-layer already trained Convolutional Neural Network has also been enhanced by using a transferable learning approach to categorize ultrasound pictures of polycystic ovarian tissue. They compared each of the approaches using quantitative performance assessment measures and determined that PCONet performed better than each of them, obtaining 98.12% correctness on sample pictures, whereas Inception III required minor adjustments with a 96.56% efficiency. Hdaib et al. [33] produced a powerful diagnostic model using MATLAB. They implemented numerous machine algorithms using 7 techniques. The outcomes showed that the linear differential classifier surpassed the K-Nearest Neighbor model based on accuracy while outperforming it in terms of sensitivity. Furthermore, the precision and accuracy of each model were compared with the accuracy and precision of 4 different articles that employed a similar Polycystic Ovary Syndrome dataset and used identical systems, testing procedures, models, and categories. According to Alagarsamy et al. [34], a heat map preprocessing phase is used in a technique to identify PCOS. After preprocessing, information is used for both testing and training.

The researchers used (SVM, NB, KNN and ensemble methods) for classification and training. SVM and KNN are a few ensemble approaches. With an f1-score of 98.06452%, the Ensemble approach performed better than other methods. To improve the amount of data collected from minorities, Inan et al. [35] presented a robust sampling strategy that integrates all oversampling and under-sampling strategies. They then used two types of decision-making procedures, the Chi-Square test for categorical features and the ANOVA test for numerical characteristics, as well as 6 various machine-learning algorithms, with the XGBoost classification model surpassing the others. Meena et al. [36] proposed a method employing Neural Fuzzy Rough Set and Artificial Neural Network approaches, and then used these minimized sets of attributes in four different types of predictive models to identify PCOS, with improved performance when compared to the different five types of traditional feature selection methods. An improved chi-squared method for feature selection was described by Nasim et al. [37], they then carried out a performance comparison analysis of 10 hyper-parametrized machine learning models for PCOS prediction. The principal component analysis, or PCA, was used by Denny et al. [38] to determine the best characteristics of their PCOS dataset, which was collected from Thrissur infertility treatment centers. A variety of classification techniques, including CART, KNN, Gaussian NB, RF, SVM, and LR, were then used in the PCOS dataset to aid in the early identification of the condition. The results showed that RF outperformed other classifiers with an accuracy rating of 89.02%. Rachana [39] uses ultrasound pictures to analyze numerous characteristics using a combination of segmentation, feature extraction, and classification techniques to identify PCOS. To improve on earlier work and create a more accurate model, several segmentation and classification methods were researched. KNN is used as a classification method.

In their research, Neto et al. [40] evaluated the effectiveness of many classifiers, including SVM, MLP, RF, LR, and Gaussian NB, in the classification of the PCOS Kaggle dataset. RF might outperform other classifiers by reaching a 95% accuracy rate and a 96% precision rate. Using the Kaggle PCOS dataset, which consists of nine characteristics, Munjal et al. [41] employed three machine learning techniques—Extra Trees, RF, and Decision Trees—as well as Genetic Algorithms (GA) as a feature selection approach to diagnose PCOS. With the best accuracy rate (88%) among the two classifiers, the results indicated that Extra Trees may exceed both. Zhang and colleagues [42] employed XGB, KNN, RF, and a stacking classifier model that included XGB, RF, and KNN in its first layer and XGB in its second layer to classify two datasets derived from the Sheng Jing Hospital's Raman spectra at China Medical University. Two sets of data from the plasma samples and the follicular fluid samples were also included in the dataset. According to experimental data, the stacking classifier model was capable of surpassing other classifiers and reached a better accuracy rate (89.32%) while using follicular fluid rather than plasma samples. The main objective of P Soni, et al. [43] is to improve medical imaging's ability to diagnose polycystic ovary syndrome.

By utilizing an identification approach, the suggested methodology obtained notable performance metrics, such as an accuracy of 86.3% and an F1 Score of 87%. Detecting tiny cysts in female patients is the main area that must be improved. Variable extraction and classification are the two essential processes that make up the framework of the technique. Particularly in polycystic ovary identification, these results could enhance the development of medical image processing methods. Pushkarini and Anusuya [44] classified a PCOS dataset that was gathered from an infertility clinic and research Centre using RF, Linear Regression, and K-Nearest Network (KNN) techniques. According to experimental findings, RF outperformed the other two approaches in terms of R2. A hybrid identification strategy (combining ANNs and NB) was used by Thomas and Kavitha [45] to categorize a clinical PCOS dataset that was collected from Thodupuzha's hospitals and scanning centers. Additionally, their hybrid classifier achieved an accuracy rate of 95%. Tanwani [46] classified the Kaggle PCOS dataset using LR and KNN after applying Felder techniques to find a connection between its characteristics. According to the results, LR outperformed KNN in terms of accuracy rate by 92% (using 10 characteristics of choice). Islam et al. [24], the goal was to identify PCOS using ovarian ultrasonography. They used a CNN (VGG-16) and ensemble models (XGBoost) and achieved an impressive 99.89% accuracy.

The DNN method for feature extraction outperformed others in accuracy, precision, recall, and F1 score, highlighting the effectiveness of stacking ensemble machine learning for PCOS detection. Several machine learning algorithms were employed by Kinjal Raut et al. [47] to identify PCOS, with the CatBoost Classifier obtaining the best accuracy at 92.64%. With an accuracy of 82.79%, Decision Tree was the most accurate method, followed by SVC (69.05%), Random Forest (89.42%), Logistic Regression (83.32%), K-Nearest Neighbors (74.34%), XBRF (85.89%), and Cross-Validation (85.69%). The iterative method was used by Vikas et al. [48]. Accuracy is evaluated in each iteration against its predecessor. The models that are employed are the hyperparameter-tuned transfer learning model, a model with extra hidden layers and a dropout layer, and the standard CNN (benchmark) model. Compared to the original model, the accuracy increased by 10%0.94% was the maximum accuracy that was attained. Kumari et al. [49] employ the GAN (Generative Adversarial Network) design to create simulated pictures for higher accuracy using VGG-19, DenseNet-121, ResNet-50, and Inception V3 and model stacking. Among them, VGG-19 achieves the best accuracy with improved sensitivity and specificity, i.e., around 70%. The model included DenseNet-121, ResNet-50, Inception V3, and model stacking.

Because there is a limited amount of dataset, data augmentation and a synthetic picture generator approach are applied. Using a clinical dataset from Kottarathil's Kaggle repository, Tiwari et al. [50] assess machine learning techniques for PCOS diagnosis. Different approaches are assessed using non-invasive screening measures. With an accuracy of 93.25%, the Random Forest approach performs better than other techniques. The prediction outcomes of RF is analyzed using out-of-bag error. A dataset of 541 individuals with 45 illness variables was used by Lakshmi et al. [51] from the UCI ML repository. Researchers make use of ensemble learning algorithms such as XGBOOST, AdaBoosting, RF, and Bagging classifiers. Having a 91.7% accuracy and an F1 score of 92%, the model accurately predicted PCOD in young women. Using optimized machine learning classifiers, KVora et al. [52] provide a technique for early PCOS detection. Using Pearson's correlation coefficient, they carried out a thorough EDA on the dataset and chose six characteristics. To overcome the class disparity, excessive sampling was also employed. The maximum prediction accuracy was obtained by optimizing classifiers using the optimal hyperparameters. With a prediction accuracy of 0.995, the Optimized KNN fared better than 11 other classifiers. An approach for predicting premenstrual syndrome (PCOS) based on minimum parameters, such as weight, BMI, duration of the cycle, and hormone levels, was established by Saha et al. [53]. Using five machine learning algorithms, they evaluated the procedure on a dataset of 541 patients. The models with the best accuracy for predicting PCOS were Random Forest and Support Vector Machine. By encouraging those who are in danger to seek medical assistance, the system may help with early detection. 30% of the dataset is utilized for testing and 70% of it is used for training. Four machine learning classifiers were combined to create a unique stack model that was suggested and had an accuracy of 90%. The models' respective accuracy percentages for the testing data were 88%, 91%, 90%, 69%, 86%, and 90%. Kaur et al. [54] examined several techniques, including DT, SVC, RF, LR, K-Nearest Neighbor, CatBoost and Gradient Boost techniques, for identifying PCOS in patient records. With a 99.8% accuracy rate, the CatBoost Classifier provided the greatest results and showed promise for a successful outcome.

To properly predict a disease, Sinthia et al. [55] integrated many data mining techniques and emphasized the significance of characteristics. SVM demonstrated the best accuracy at 91%, followed by KNN at 75%, K-means at 72%, and linear regression at 85%. They also utilized K-means, KNN, and K-means. A beneficial approach for follicle recognition in ultrasound images is suggested by Kumar et al. [56]. Four methods are used in the process: PSO, CPSO, PIO and GPI optimization. Image preprocessing and follicle extraction are carried out following segmentation. The effectiveness of these techniques is compared in the research between hand segmentation using Dice and Jaccard coefficients, similarity measurements, and MATLAB. With an accuracy range of 89.7569% to 90.6515% with four test inputs, the Gaussian pigeon-inspired technique outperforms the current region-based active contour approach, which has an accuracy of 83%. The study finds that a more accurate approach to follicle detection is provided by the algorithm influenced by Gaussian pigeons. Using ELM and Bayesian optimization, Swapnarekha et al. [57] created a prediction model for PCOS early identification. Class imbalance issues were resolved by using random oversampling. To choose the ideal hyperparameters, Bayesian optimization was applied. The model was tested against various machine-learning methods are evaluated on a Kaggle PCOS dataset. Comparing ELM and Bayesian optimization against alternative techniques, the findings demonstrated that they attained an accuracy rate of 99.31%. To detect and diagnose PCOS with the minimum possible factors, Prajna et al. [58] employ ML techniques. Four techniques are utilized by the model: Chi-Square, LR, DT, and RF classifier.

The dataset utilized in the model comes from the "KAGGLE" database. With its best accuracy, the Random Forest classifier (Chi-Square) assists women with PCOS by increasing their quality of life through early identification, prognosis, and treatment recommendations. Combining the XGBoosting, RF, and SVM algorithms, Swamy et al. [59] created a hybrid machine-learning model. They compared many methods, such as GB, LR Rf, AdaBoosting, DT, Support Vector Machine and MLP, and evaluated the model on a PCOS dataset. When it came to early detection of PCOS, the hybrid model had a 93.8% accuracy rate. Rathod et al. [60] developed a machine learning system for diagnosing and predicting the outcome of PCOS treatment using seven classifiers: CatBoost, SVM, LR, RF, DT Bernoulli Naive Bayes, and K-Nearest Neighbor. After comparing the precision. accuracy, sensitivity, F1 score, and AUC, researchers concluded that CatBoost was the most accurate in determining if a woman with PCOS should visit a doctor. Using machine learning methods, Adla et al. [61] studied how to create a model for PCOS diagnosis. An approach known as hybrid feature selection was employed to preprocess and minimize a dataset including 39 characteristics from 541 individuals.

The SVM with a Linear kernel (Linear SVM) was the most accurate and precise classification technique after many others were trained and assessed. Outperforming other models in terms of accuracy and precision, the model obtained a recall of 80.6%. Katarya et al. [62] have created a novel automated diagnosis method for PCOS prediction based on clinical data and machine learning. This system outperforms earlier diagnostic techniques with an astounding 90.74% accuracy. Praba et al. [63] suggested an automated methodology to categorize women with PCOS and those without it using machine learning techniques. By applying effective feature selection techniques and Raman spectra, they pre-processed follicular fluid samples from 100 women. Analysis was done on the effectiveness of sophisticated machine learning classifiers such as Random Forest, Ada Boost, Multilayer Perceptron, and decision trees. Researchers have shown that utilizing follicular fluid samples, the Raman spectroscopic model can predict PCOS with 100% accuracy. Random forest, decision tree, naive bayes, logistic regression, and artificial neural network are the five DM approaches that Marreiros et al. [64] used to create a model to detect PCOS patients. The model with the highest accuracy, precision, sensitivity, and specificity scores was the RF classifier, which had a score of 93.06%.

The accuracy of the model was 1.9% lower than in a prior investigation that employed feature selection based only on correlation weight Ahmetašević et al. [65] created a method that uses ANN to classify cases of polycystic ovary syndrome. The system, which included 1000 healthy subjects and those with disease, achieved an accuracy of 96.1%, sensitivity of 96.8%, and specificity of 90%, indicating the system's significant potential in this field. The system predicted 157 positive cases and 23 negative cases. An enhanced active contour approach without an edge was employed by Kumar et al. [66] to quickly segment data and find tiny follicles. The experimental findings show that, in comparison to the current approach, the enhanced algorithm may obtain the desired result with fewer computing steps and iterations. To produce more convergent clusters, Setiawati et al. [67] propose an image clustering technique for follicle segmentation using the method known as PSO with a freshly modified nonparametric fitness function that uses the Mean Structural Similarity Index (MSSIM) and Normalized Mean Square Error (NMSE).

Compared to PSO image clustering without improved contrast, PSO image clustering with contrast enhancement results in more intra-cluster, intracluster, and quantization errors. During segmentation, Rihana et al. [68] used both horizontal and vertical scanline thresholding, and for minimizing noise along with contrast enhancement, they employed the multi-scale morphological approach. The primary, secondary, area, and perimeter axes are the factors that were considered throughout segmentation. Sitheswaran et al. [69] suggested an improved labelled watershed approach that uses object growth to locate follicles and a median filter to eliminate speckle noise. To denoise the picture in the wavelet domain, Raj [70] employed a thresholding method. The use of contrast enhancement makes use of a morphological method. The picture is subjected to the fuzzy c-means clustering technique for follicle segmentation, which is faster than the current approach.

Table 1 lists the objectives of the works that have already been done, the algorithms that have been used to achieve those goals, the flaws in those works, and the authors' responses. Researchers may quickly access all the material as this table provides an overview of all papers. These remarks will also assist researchers in conducting their work successfully. Several important points may be obtained from the debate above. Traditional PCOS diagnostic techniques were not fully described in the papers that were already published. The techniques have been addressed in detail. The researchers only discussed the datasets they utilized, without properly discussing any other existing datasets. They were unable to determine how the set of data performed in various machine learning (ML) techniques. Additionally, Researchers did not evaluate how well other literature performed in comparison to their own. However, the obstacles and potential opportunities for utilizing ML to diagnose Polycystic Ovary Syndrome are not clearly stated.

The goals, employed methods, and important observations of the PCOS detection efforts that have already been done are summarized in Table 1.

3 Techniques for Recent PCOS Detection

This section describes many PCOS methods for identification, along with their parameters and structural details. Parent detection approaches for PCOS are divided into two groups. The first technique is the conventional identification approach, and the second uses Machine Learning based identification methods. The hormonal range for PCOS and normal are crucial in traditional screening methods. This section includes a table with a full explanation of this hormonal range for your convenience. The machine learning (ML) algorithm structures are provided for the researchers' easy comprehension. There is a diagram that includes all the PCOS detection methods. Additionally, a full illustration of how PCOS is usually identified by doctors using conventional techniques is provided.

3.1 Traditional Methods

Several hormone levels are considered while determining PCOS.

- Luteinizing hormone: The female body needs the chemicals luteinizing hormone (LH) and a hormone that stimulates follicles (FSH). Before 24 h of ovulation, LH levels increase to around 25–40 mlU/ml. LH levels in PCOS-afflicted women are frequently around 18 mIU/ml while FSH levels are about 6 mIU/ml. This used to be considered a crucial element in PCOS treatment.
- Testosterone: The overall amount of testosterone in a person's body, comprising testosterone that is not synthesized, is known as testosterone. The range is from 6.0 to 86 ng/dl. Free testosterone is the amount of active, male hormones that are liberated in the human body, which typically ranges from 0.7 to 3.6 pg/ml. Increased levels of masculine hormones, both full along free prevalent in Polycystic Ovary Syndrome females.
- DHEA-S: Most PCOS women have levels that are higher than 200 ug/dl.

Table 1 Summary of recent research studies on PCOS detection techniques

References	Goal	Used techniques	Observations
Chitra et al. [21]	To automatically identify PCOS	Convolutional neural network	The watershed segment algorithm is utilized and provides good accuracy Because CNN is highly fast, it may be utilized for classification
Hassan and Mirza [22]	Classify the Kaggle PCOS data set using ML models	RF, SVM, CART, Naive Bayes (NB), and LR	Testing data showed that RF might beat other classifiers by obtaining a 96% classification accuracy rate
Cahyono et al. [23]	To diagnose PCOS and strengthen the system's ability to handle real-world data	CNN	The highest performance is shown in this study when learning rate 10 increased to power 6 combined with a dropout of 0.1. nevertheless, several combinations can be tested for improved performance
Islam et al. [24]	Identify PCOS with ovarian ultrasonography	CNN(VGG-16), ensemble models (XGBoost)	They identified PCOS using Ovarian Ultrasonography achieving 99.89%, accuracy with CNN(VGG-16) and XGBoost ensemble
Mehr and Polat [25]	Early detection of PCOS	Ensemble RF, MLP, AdaBoost and extra tree	Not mentioned
Gupta et al. [26]	Conducted discriminant study with linear and quadratic forms for binary classification	Quadratic discriminant analysis, ensemble ML models	Achieved training accuracy of 97.37% and testing accuracy of 95.92%. Identifies an area for improvement in analysis Proposes a modified ensemble method for PCOS diagnosis
Vasavi et al. [27]	To develop an application that records women's menstrual symptoms	DT, LR, SVM, KNN, NB and RF	The suggested work created an app to support the lifestyle in addition to detecting PCOS. However, it would have been preferable if the feature selection process had been explained in detail
Deshpande and Wamankar [28]	Automatic PCOS feature extraction utilizing a multiscale morphological method	SVM	The technique of canny edge detection is employed for segmentation. It takes a lot of work, but it yields flawless edges. The Sobel operator is a time-efficient tool, although its edges are not very smooth

(continued)

Table 1 (continued)

References	Goal	Used techniques	Observations
Khanna et al. [29]	To help in the decision-making of medical experts by identifying thyroid and PCOS	AdaBoost, RF, NB, DT, LR, KNN, STACK 1, STACK 2, STACK 3, XGBoost and Extra Trees	The model is performing well when the AUC curve almost reaches the highest point
Dutta et al. [30]	To determine which ML classification technique is the best	DT, SVM, LR, KNN and RF	SMOTE is used to balance data sets; the enhanced SMOTE ADASVN version can be applied as it improves sample reality
Prapty et al. [32]	Create a PCONet CNN-based model for detecting PCO in ovarian images from ultrasounds	PCONet, InceptionV3	PCONet is more effective for the task, demonstrating superior performance in identifying polycystic ovarian tissue. InceptionV3, though capable, needed adjustments to reach competitive accuracy
Hdaib et al. [33]	To develop a novel, simple approach for PCOS diagnosis	LDA, LR, DT, SVM, NB, KNN, NN	The linear discriminant classifier demonstrated superior performance over KNN, indicating its effectiveness for PCOS diagnosis. The broader comparison provides context by evaluating the model against similar studies, reinforcing its significance
Alagarsamy et al. [34]	PCOS fertility and infertility prediction	Ensemble (SVM + KNN), KNN, NB and SVM	SVM and NB have comparatively low error rates Combining these two models will improve performance
Inan et al. [35]	Optimized feature selection and sampling to enable high-voltage gradient boosting for PCOS identification	SMOTE, XGBoost, ENN, ANOVA test chi-square test	The study suggests a sampling technique to enhance minority samples and eliminate outliers from medical datasets that combine oversampling and under-sampling
Meena et al. [36]	Develop an approach for detecting PCOS (Polycystic Ovary Syndrome)	NFRS, ANN	The combination of NFRS and ANN techniques, along with the reduced set of features, proved effective in enhancing the performance of classification models for detecting PCOS. The comparison with traditional feature selection methods highlights the superiority of the proposed approach

(continued)

Table 1 (continued)

References	Goal	Used techniques	Observations
Nasim et al. [37]	Create an improved (CS-PCOS) approach for selecting features and compare the performance of 10 hyper-parametrized ML systems for PCOS identification	(CS-PCOS)	The improved CS-PCOS method serves as an advanced approach for feature selection, enhancing the performance of machine learning models in predicting PCOS. The analysis of hyper-parametrized models provides valuable insights for model selection in this context
Denny et al. [38]	The suggested approach uses the best scientific and metabolic indicators to anticipate and detect PCOS early	NB, LR, KNN, CART, RF, SVM	By exploiting optimal scientific knowledge, the suggested method strives for early detection and prediction of PCOS. The use of PCA for feature transformation allows for improved efficiency, and the comparison of various ML techniques provides insights into their effectiveness for PCOS classification
Rachana [39]	Identify PCOS (Polycystic Ovary Syndrome) by analyzing several features in ultrasound images	KNN	The research focus on improved methods indicates a commitment to advancing the state-of-the-art in PCOS detection
Neto et al. [40]	Evaluation of the PCOS Kaggle dataset's categorization using several classifiers	SVM MLP, RF LR, NB	This research focuses on assessing the performance of multiple classifiers on the PCOS Kaggle dataset. The standout performance of Random Forest, with high accuracy and precision rates, underscores its efficacy in handling the dataset for PCOS classification
Munjal et al. [41]	Detect PCOS by utilizing the Kaggle PCOS dataset	Extra trees, RF and DT	This research focuses on leveraging machine learning techniques and genetic algorithms for effective PCOS diagnosis. The preference for Extra Trees, indicated by the best accuracy rate, highlights its suitability for the diagnostic task in comparison to other classifiers

(continued)

Table 1 (continued)

References	Goal	Used techniques	Observations
Zhang [42]	Classify datasets using different techniques	XGB, KNN, RF	This research aimed to classify Raman spectra datasets from Shengjing Hospital, considering different sample types. The stacking classifier model demonstrated superior performance, particularly with follicular fluid samples, highlighting its potential for accurate classification in this specific context
Soni et al. [43]	Polycystic ovary syndrome detection	Classification approach	This study improves the classification-based detection of polycystic ovarian disease, achieving an F1 Score of 87% and an Accuracy of 86.3%. The primary focus for detection in females is small cysts. Two processes are involved to enhance medical image processing for polycystic ovary identification: variable categorization and extraction
Pushkarini and Anusuya [44]	Determine the PCOS status of a dataset collected from an infertility clinic and research facility	RF, LR, KNN	By utilizing several methods, the authors attempted to categorize a PCOS dataset. The outcome of the studies reveals that RF is more successful in terms of R2 than LR and KNN
Thomas and Kavitha [45]	Identify a diagnostic PCOS dataset taken from Thodupuzha's clinics and scanning centres using a hybrid identification method	NB, ANN	To classify a clinical PCOS dataset, the scientists employed a hybrid technique that included ANNs and NB. The impressive 95% accuracy rate demonstrates how well their hybrid classifier identified PCOS instances in the sample

(continued)

Table 1 (continued)

References	Goal	Used techniques	Observations
Tanwani [46]	To determine the best-supervised classification model for detecting Polycystic ovary syndrome using ML approaches	ANN, LR	The purpose of this research is to categorize the Kaggle PCOS dataset. Employing both LR and KNN. The application of Felder techniques enhances the understanding of connections between dataset characteristics. The significant finding is the superior accuracy rate of LR compared to KNN, showcasing its effectiveness in the specific context of PCOS dataset classification
Raut et al. [24]	Identify PCOS	Cat Boost, DT, SVC, RF, LG, KNN, XBRF, cross validation	Cat Boost achieved the highest accuracy at 92.64%. DT was the most accurate method at 82.79%
Raut et al. [47]	Improve accuracy through an iterative process	Basic CNN, CNN with more layers, Transfer learning with hyperparameter tuning	Achieved 10% improvement from the initial model, reaching the highest of 94%
Vikas et al. [48]	Enhance accuracy with GAN design	Model Stacking, VGG-19, DensetNet-121, ResNet-50, Inception V3	VGG-16 achieved the best accuracy around at 70% with improved sensitivity and specificity. Data augmentation and a synthetic picture generator were applied due to the limited dataset
Kumari et al. [49]	Simulated picture creation, and improved accuracy	Inception V3, Model Stacking, GAN, VGG-19, DenseNet-121, ResNet-50	VGG-19 achieved the best accuracy at around 70% with improved sensitivity and specificity. Data augmentation and synthetic picture generators were used due to a limited dataset accuracy, aiding in early identification and treatment recommendations
Tiwari et al. [50]	Assess PCOS diagnosis using clinical dataset	RF	The Random Forest algorithm achieved 93.25% accuracy, surpassing other algorithms and its prediction performance was evaluated using the out-of-bag error
Lakshmi et al. [51]	Predict PCOD in young women	GB, AdaBoost, RF, bagging classifier	

(continued)

Table 1 (continued)

References	Goal	Used techniques	Observations
Vora et al. [52]	Early PCOS detection using optimized classifiers	Optimized KNN and 11 other classifiers	Attained 99.5% prediction, accuracy, leveraging Pearson's correlation coefficient for feature selection and mitigating class disparity through excessive sampling
Saha et al. [53]	Predict PCOS using minimal parameters	RF, SVM, Stack model	The models with the highest accuracy were RF and SVM The stack model, combining 4 classifiers achieved 90% accuracy. Testing data accuracy ranged from 69 to 91%
Kaur et al. [54]	Identify PCOS in patient records	DT, SVC, RF, LR, KNN, XGBRF, CATBoost classifier	The Cat Boost classifier outperformed others with a remarkable 99.89% accuracy, showing promise for successful outcomes
Sinthia et al. [55]	Predict a disease using data mining techniques	SVM, KNN, L-means, LR	Integrated Various data mining techniques for disease prediction. SVM exhibited the best accuracy at 91%, followed by KNN at 75%, K-means at 72% and Linear Regression at 85%
Kumar et al.. [56]	Follicle recognition in ultrasound images	PSO, CPC, PIO, GPIO	Proposed a beneficial approach for follicle recognition using various optimization methods Gaussian Pigeon Inspired Optimization outperformed the current approach with an accuracy range of 89.76% to 90.65%
Swapnarekha et al. [57]	PCOS early identification model	ELM, Bayesian optimization	Created a PCOS early identification model using ELM and Bayesian optimization. Resolved class imbalance with random oversampling. Achieved an impressive accuracy rate of 99.31%
Prajna et al. [58]	Detect and diagnose PCOS with minimal factors	Chi-square, LR, DT classifier, RF classifier (chi-square)	Employed machine learning techniques for PCOS detection using a KAGGLE database. RF classifier (chi-square) achieved the best

(continued)

Table 1 (continued)

References	Goal	Used techniques	Observations
Swamy et al. [59]	Early detection of PCOS	XGBoosting, RF, SVM	Created a hybrid machine learning model using XGBoosting, RF and SVM. Achieved 93.8% accuracy for early detection of PCOS
Rathod et al. [60]	PCOS treatment can be predicted and diagnosed early	CatBoost, RF, SVM, LR, Bernoulli Naïve Bayes, DT, k-Nearest Neighbor	Developed an ML system for early PCOS diagnosis and therapy treatment. CatBoost exhibited the highest accuracy and recommended visiting a doctor
Adla et al. [61]	PCOS diagnosis using machine learning	SVM with linear kernel	A combined selection of features technique was used to develop a method for PCOS detection. The SVM with a linear Kernel (Linear SVM) achieved high accuracy
Kataraya et al. [62]	Automated PCOS prediction	Machine learning	Developed a novel method with 90.74% accuracy
Praba et al. [63]	Automated PCOS categorization using Raman Spectroscopy	RF, AdaBoost, MLP, decision tree	Achieved 100% accuracy with follicular fluid samples
Marreiro et al. [64]	PCOS detection model development	RF, DT, NB,LR,ANN	RF classifier achieved the highest accuracy 93.06%,1.9% lower than a prior investigation
Ahmetasevic [65]	PCOS classification using ANNs	ANNs	Developed a system achieving 96.8% sensitivity and 90% specificity. Predicted 157 positive and 23 negative cases
Kumar et al. [66]	Efficient follicle segmentation using enhanced active	Enhanced active counter	The improved algorithm achieves the desired results with fewer computing steps and iterations
Setiawati et al. [67]	Image clustering for follicle segmentation using PSO with modified fitness function	PSO with contrast enhancement	Improved contrast in PSO image clustering reduces intra-cluster errors and enhances segmentation results
Rihana et al. [68]	Segmentation using horizontal and vertical scanline thresholding	Multi-Scale morphological approach	Utilized multi-scale morphological approach with horizontal and vertical scanline thresholding for segmentation

(continued)

Table 1 (continued)

References	Goal	Used techniques	Observations
Sitheswaran et al. [69]	Improved labelled watershed approach for follicle detection	Median filter, watershed	Implemented improved labelled watershed approach with object growth for follicle detection and used a median filter to remove speckle noise
Raj et al. [70]	Denoise image in the wave	Thresholding, morphological method, Fuzzy C-means clustering	Employed thresholding for denoising in the wavelet domain, contrast enhancement through a morphological method and utilized Fuzzy C-means clustering for faster follicle segmentation

- Prolactin: Between 25 and 40 ng/ml of prolactin is frequently seen in women with PCOS.
- Estrogen: Some PCOS ladies have estrogen levels that fall between the normal range (25–75 pg/ml).
- TSH: In PCOS women, the typical range for Thyroid Stimulating Hormone is 0.4–3.8 uIU/ml [71].

Table 2 depicts hormonal fall between healthy and Polycystic Ovary Syndrome—affected individuals. Researchers can gain a greater understanding of conventional PCOS diagnostic techniques by using this table.

For manually identifying PCOS, doctors examine the symptoms and the number of hormones. The most expensive and time-consuming method of identifying PCOS is a hormone test. To identify the signs of PCOS, a battery of survey questions is often employed. Surveys might be used to find relatives of PCOS patients who have clinically apparent PCOS. Even though a written questionnaire can identify a large percentage of affected moms, around 50% of affected sisters go unreported [72].

Using information from the ultrasound scan on the number of cysts, doctors can manually identify PCOS. PCOS may be present if each ovary has more than 12 follicles measuring 29 mm [73]. Doctors identify PCOS and personally count the

Table 2 Normal range, PCOS-related hormonal changes, and the Detection method used to identify them

Feature	Normal range	Range in PCOS	Method
FSH	4.27–21.5 IU/L	Increased level	Blood test
LH	5–25 IU/L	Increased level	Blood test
Prolactin	Less than 25 ng/mL	Normal	Blood test
DHEA	35–430 ug/dl	Greater than 200	Blood test
Testoterone	15–70 ug/dl	Increased level. Mostly less than 150 ng dL	Blood test
Estrogen	30–40 pg/mL	Level falls	Blood test

cysts. It takes a lot of time. Additionally, mistakes can happen, such as counting incorrectly, mistakenly overlooking a cyst, mistaking a bump for a cyst, etc. There is another ultrasound available besides the one of the abdomens. It is known as the transvaginal USG. A greased probe is placed into the female patient's vaginal canal, and the doctor monitors the internal state of the organs or investigates the internal workings. They carefully identify each follicle inside the ovaries and assess the condition of the uterus and cervix. To identify PCOS, they also evaluate ovarian volume. However, this procedure is not faultless. When calculating the volume of the ovaries by counting the follicles within, medical professionals are susceptible to error. This approach is not easy, and many individuals dislike it due to barriers of culture and society.

3.2 Machine Learning

The goal of Machine Learning (ML), a discipline that is expanding, is to emulate the intelligence of people by learning from their environment [74]. Clustering, as well as classification, are the two fundamental subfields of machine learning. Classification is a supervised data mining process. Using this method, a new observation is categorized into pre-existing categories or classes based on its structure. These categories are identified via a classification model, allowing us to determine the set of Unnoticed class identifiers or classification labels of information with unidentified labels [75]. Following list contains the classification algorithmic methods:

Support Vector Machine (SVM): It is a regression as well as classification prediction technique that uses machine learning (ML) theory to increase the accuracy of predictions while simultaneously preventing erroneous information from fitting. Support Vector Machine is a useful method for classifying data. SVM develops a hyperlink that divides data into many classifications. It can resolve both linear and nonlinear issues. It may also be used to solve challenges in everyday life. The fundamental benefit of SVM is that training is relatively easy. There isn't a local ideal as there is in neural networks. The need for a strong kernel function [76] is one limitation. According to Nandipati et al. [77], In terms of accuracy, recall, and f1-score, the SMOTE-based Support Vector Machine succeeded well on average.

Naïve Bayes (NB) uses supervised learning and is based on the Bayes theorem. The assumption that characteristics are independent of class allows this classifier to significantly increase learning. Each set of features being categorized by this algorithm is independent of the others. Even though independence is often a weak criterion, NB typically outperforms more complex classifiers in practice [78]. It was employed by Prapty and Shitu [32], who got 93% accuracy out of it, which was second only to RF in terms of accuracy. K-Nearest Neighbors (KNN) is a simple yet efficient classification technique. It is a form of guided learning technique that categorizes data based on proximity. It operates by measuring the distance between a fixed location and other nearby sites.

Non-parametric classification techniques like the KNN classification method are simple but useful in a variety of situations. It is necessary to identify a data record's "k" nearest neighbours to categorize it. A neighbourhood of "t" is created when neighbours are found. The categorization of t is normally determined by the neighbourhoods' data records, with or without taking into account distance-based weighting. However, a suitable value for k must be chosen to employ KNN. The KNN technique is impacted by k in certain ways [79]. Decision tree (DT) classification technique is simple yet effective. Giving humans easily understood categorization criteria is one of its important advantages. The DT has the drawback that all quantitative properties must be sorted before a node can be separated. This takes a lot of time and memory, especially while developing decision trees on enormous amounts of data [80] The DT's operational mechanism is shown in Fig. 5. This diagram illustrates how the categorization process is carried out by the decision node. Although DT has several drawbacks, many researchers still employ it because of its potent potential. Aggarwal and Pandey [81] used DT in conjunction with other categorization techniques. The bulk of contemporary operations make use of DT.

Random Forest (RF) has been extremely successful as an all-around categorization and prediction tool. Whenever the total amount of components exceeds the number of evaluations, the technique that takes the estimates of numerous arbitrary DTs outperforms. Figure 8 shows the inner workings of a decision tree. It is suitable for a wide range of ad hoc acquiring assignments. The root node is where the process begins. Decision nodes create classes and leaf nodes. It is ideal for large-scale difficulties due to its measures of changing importance [82]. RF was also employed in a hybrid approach that included LR. When it came to their task, this model did the best. Logistic Regression (LR) uses classification to estimate binary values rather than a regression approach. By matching data to a logit function, the binary values and the likelihood that an event will occur are predicted for the given set of independent variables. The probability prediction is done, and the results' values range from 0 to 1 [22]. In cases of PCOS detection, LR is used at a high rate. RF and it are combined to create hybrid models. In terms of PCOS identification, this hybrid model performs better than the alternative categorization.

Clustering is a data collection technique used in machine learning (ML). A collection of points of information may be classed according to some criteria and common qualities. This plan of action is useful for finding abnormalities. This is a technique for obtaining without supervision information. [75]. In K-means, unidentified information is divided into categories using this clustering approach. K indicates the number of pre-clusters that should be produced throughout processing; for example, if K is 2, two clusters are created; if K is 3, a total of three clusters are created; consequently.

To understand how the human brain processes information, ANN analyses the neural network, creates a basic model, and assembles several networks using different types of connections. A network is a representation of how information is processed and is made up of many linked nodes (or neurons). Every single node indicates a unique function of activation, which is a particular outcome value. Every two nodes have a link that indicates the transmitted signal's strength as it moves via the link, which is equal to an ANN's memory. The network's output varies based on its connectivity, weight amount, and reward function [84]. CNNs were the most basic kind of deep neural network (DNN). Between two numbers, it is a linear mathematical process. There are several layers in CNN, including convolutional, nonlinear, pooling, and completely linked, convolutional, and wholly coupled levels; pooling and nonlinearity layers lack parameters. Non-linearity and pooling layers don't have parameters. Because of its variety, CNN performs quite well in ML [85]. When it comes to cyst detection from ultrasound pictures, it is a highly effective method. CNN-trained methods can correctly determine whether an ovary is PCOS-infected or normal. Cahyono et al. [23] detected PCOS using CNN architecture. This model successfully performed at a rate of 100%.

The input–output alignment of Multilayer Perception neural network is nonlinear. An MLP is made up of layers that provide input and output, a few layers that are hidden, and multiple hidden layers that are piled on preceding one another. Neurons in an MLP can utilize any random activation function, in contrast to neurons in a perceptron, which must employ ReLU or sigmoid stimulation processes that enforce thresholds [88]. PCOS detection at a minimum uses ANN technology.

4 Future Directions and Challenges

This section assists researchers in considering in which to focus their attention by highlighting Previous investigations that identified most of the significant issues and concerns linked with PCOS identification. In accordance with data privacy and security issues, the book chapter investigates sophisticated machine learning and deep learning models for PCOS diagnosis. The bulk of the examined research publications make use of open-source datasets, emphasizing the significance of suppressing private medical records such as ultrasound pictures, hormone profiles, genetic data, and patient information. These datasets are compliant with international requirements such as HIPAA (Health Insurance Portability and Accountability Act) and GDPR (General Data Protection Regulation), which ensure patient data is safe and secure. This review paper investigates multiple machine learning models for PCOS diagnosis using publicly available datasets. While the review emphasizes the need of patient agreement and openness in enabling the practical application of AI in healthcare, it is vital to note that patient consent does not apply to this study because no datasets are explicitly utilized in the paper. Future developments are also explained. The associated challenges for this task along with possible solutions are discussed. Researchers will be supported in identifying study gaps and carrying out follow-up work.

Regarding the existence of a few useful data sets, there are certain drawbacks. For instance, the datasets are not varied, and the datasets available for PCOS identification are rather tiny. Many of the datasets are unique. The custom datasets are minimal. However, Kaggle has a relatively limited selection of datasets. Due to the model's ability to learn and extract features, machine learning performs best when the training and test dataset is large. It is required to employ an enormous data set that is impartial to a certain geographic region and broad in scope. Additionally, a dataset must contain women of different ages to represent multiple generations. The test result won't be accurate if the dataset isn't substantial and consistent.

For every class, an evenly distributed dataset contains the same kinds of information. The current datasets are useful but unbalanced. One class has many observations, whereas the other class has few observations. The outcome is greatly impacted by this trait. Because of this, no absolute impact is discovered. The dataset may be balanced using a variety of preprocessing methods, such as PCA and SMOTE. Images must be sufficiently clear to run the CNN model, however, ultrasound images can contain a variety of sounds like salt and pepper noise, speckle noise, and other disturbances. For the best outcome, these sounds must be eliminated. The accuracy rate for some CNN models is poor because noise is not eliminated. Cyst identification is not always accurate. However, most currently published studies do not eliminate image noise. As a result, techniques like dilation, grayscale, etc., are required for noise reduction.

To make it possible for patients to accept machine-based identification over manual detection, researchers need to achieve a detection rate of 100% if they want to automate PCOS screening and use Machine Learning driven approaches within the health system. However, the rate of detection in the earlier investigations is imperfect. Most detection rates are lower than 98%. The model must be trained more and more to achieve this 2% accuracy. The area of computer vision has seen a radical transformation thanks to object detection techniques. This method can simultaneously classify and locate objects. Consequently, object detection techniques are quicker than conventional methods. Consequently, it will be highly helpful to use object detection algorithms to identify PCOS. The available literature mostly enhances classifier-based methods. However, in this field, object identification algorithms like YOLO, and Fast RCNN, or may be utilized to identify cysts in ultrasound pictures. The k-means method is the simple method of clustering utilized to identify PCOS. Numerous clustering techniques are available, including, Balanced iterative Reducing and Clustering using Hierarchies, Ordering Points to Identify Clustering Structure, Density-Based spatial clustering of applications with noise and the Gaussian mixture model. These techniques aren't utilized, nevertheless. To understand how clustering-based models identify PCOS, these methods have to be utilized. For this reason, there may be greater success with alternative clustering techniques than with K-means.

5 Conclusion

This chapter concentrated on ML and provided a summary and conceptual assessment of all current PCOS diagnostic approaches. This chapter provided an overview of existing algorithms, including their characteristics, efficacy, analysis techniques, and outcomes. The study analyzed the inadequacies and risks associated with current algorithms. Despite much study on developing a valid PCOS diagnostic paradigm, some difficulties remain unresolved. This study highlighted limitations such as a small number of collections of data, an unbalanced dataset, a low identification rate, the lack of further clustering algorithms, etc. In our forthcoming research, we want to use a larger dataset with more balanced data. Additionally, there is a lot of CNN-based optimization to be done. Additionally, there is a lot of CNN-based optimization to be done. Authors will be capable of creating innovative methods for this issue by recognizing and examining the gaps and future directions in this domain. In future research, we want to investigate the updated techniques for PCOS identification and test them in favor of a standard data set to evaluate their performance. We desire to explore the success rate of various clustering techniques. With this method, it will be simpler to comprehend the main variations between the new and previous ML PCOS detection approaches. Authors will be capable of creating innovative methods for this issue by recognizing and examining the gaps and future directions in this domain. In future research, we want to investigate the updated techniques for PCOS identification and test them in favor of a standard data set to evaluate their performance. We desire to explore the success rate of various clustering techniques. With this method, it will be simpler to comprehend the main variations between the new and previous ML PCOS detection approaches.

References

1. https://www.healthline.com/health/polycystic-ovary-disease#what-ispcos
2. Hiremath, P.S., Tegnoor, J.R.: Follicle detection and ovarian classification in digital ultrasound images of ovaries. Adv. Breakthroughs Ultrasound Imaging **5**, 167–199 (2013)
3. Gougeon, A.: Dynamics for human growth: morphologic, dynamic and functional aspects. Ovary **2**, 25–43 (2003)
4. Pache, T.D., Wladimiroff, J.W., Hop, W.C., Fauser, B.C.: How to discriminate between normal and polycystic ovaries: transvaginal US study. Radiology **183**(2), 421–423 (1992)
5. Knochenhauer, E.S., Key, T.J., Kahsar-Miller, M., Waggoner, W., Boots, L.R., Azziz, R.: Prevalence of the polycystic ovary syndrome in unselected black and white women of the southeastern United States: a prospective study. J. Clin. Endocrinol. Metab. **83**(9), 3078–3082 (1998)
6. Wallach, E.E., Donesky, B.W., Adashi, E.Y.: Surgically induced ovulation in the polycystic ovary syndrome: wedge resection revisited in the age of laparoscopy. Fertil. Steril. **63**(3), 439–463 (1995)
7. Arentz, S., Smith, C.A., Abbott, J., Bensoussan, A.: Perceptions and experiences of lifestyle interventions in women with polycystic ovary syndrome (PCOS), as a management strategy for symptoms of PCOS. BMC Womens Health **21**(1), 1–8 (2021)
8. Watson, S.: Polycystic ovary syndrome (PCOS): symptoms, causes, and treatment (2019)
9. Nelson, V.L., Legro, R.S., Strauss, J.F., III., McAllister, J.M.: Augmented androgen production is a stable steroidogenic phenotype of propagated theca cells from polycystic ovaries. Mol. Endocrinol. **13**(6), 946–957 (1999)
10. Rotterdam ESHRE/ASRM-Sponsored PCOS Consensus Workshop Group: Revised 2003 consensus on diagnostic criteria and long-term health risks related to polycystic ovary syndrome (PCOS). Hum. Reprod. **19**(1), 41–47 (2004)
11. Faoziyah, P.N., Permana, F.P., Wirayuda, T.A.B., Wisesty, U.N.: Tamper detection and recovery of medical image watermarking using modified LSB and Huffman compression. In: 2013 Second International Conference on Informatics & Applications (ICIA), pp. 129–132. IEEE (2013)

12. Purnama, B., Hasyim, A., Septiani, M.D., Wisesty, U.N., Astuti, W.: Follicle detection on the use of images to support the determination of polycystic ovary syndrome. In: Journal of Physics: Conference Series, vol. 622, no. 1, p. 012027. IOP Publishing (2015)
13. Azziz, R.: PCOS: a diagnostic challenge. Reprod. Biomed. Online **8**(6), 644–648 (2004)
14. Raghavendra, S., Santosh, K.J.: Performance evaluation of random forest with feature selection methods in prediction of diabetes. Int. J. Electr. Comput. Eng. **10**(1), 353 (2020)
15. Wissel, T., Pfeiffer, T., Frysch, R., Knight, R.T., Chang, E.F., Hinrichs, H., Rieger, J.W., Rose, G.: Hidden Markov model and support vector machine-based decoding of finger movements using electrocorticography. J. Neural Eng. **10**(5), 056020 (2013)
16. Jyothi, D.S., Vasavi, G.: Polycystic ovary syndrome detection using various machine learning methods—A review. J. Adv. Res. Dyn. Control Syst. (2017)
17. Vaswania, J., Mulchandanib, H., Vaghelac, R., Pateld, R.: A systematic literature review on the diagnosis of PCOS using machine learning algorithms. GIT-J. Eng. Technol. (2022)
18. Pulluparambil, S.J., Bhat, S.: Medical image processing: detection and prediction of PCOS—A systematic literature review. Int. J. Health Sci. Pharmacy (IJHSP) **5**(2), 80–98 (2021)
19. Isah, O.R., Usman, A.D., Tekanyi, A.M.: A review on computer-assisted follicle detection techniques and polycystic ovarian syndrome (PCOS) diagnostic systems (2015)
20. Agrawal, A., Ambad, R., Lahoti, R., Muley, P., Pande, P.S.: Role of artificial intelligence in PCOS detection. J. Datta Meghe Inst. Med. Sci. Univ. **17**(2), 491–494 (2022)
21. Sumathi, M., Chitra, P., Prabha, R.S., Srilatha, K.: Study and detection of PCOS related diseases using CNN. In: IOP Conference Series: Materials Science and Engineering, vol. 1070, no. 1, p. 012062. IOP Publishing (2021)
22. Hassan, M.M., Mirza, T.: Comparative analysis of machine learning algorithms in diagnosis of polycystic ovarian syndrome. Int. J. Comput. Appl. **975**, 8887 (2020)
23. Cahyono, B., Mubarok, M.S., Wisesty, U.N.: An implementation of convolutional neural network on pco classification based on ultrasound image. In: 2017 5th International Conference on Information and Communication Technology (ICoIC7), pp. 1–4. IEEE (2017)
24. Suha, S.A., Islam, M.N.: An extended machine learning technique for polycystic ovary syndrome detection using ovary ultrasound image. Sci. Rep. **12**(1), 17123 (2022)
25. Danaei Mehr, H., Polat, H.: Diagnosis of polycystic ovary syndromethrough different machine learning and feature selection techniques. Heal. Technol. **12**(1), 137–150 (2022)
26. Gupta, A., Soni, H., Joshi, R., Laban, R.M.: Discriminant analysis in contrasting dimensions for polycystic ovary syndrome prognostication (2022). arXiv:2201.03029
27. Vasavi, R.R., Prathibha, S.P., Valiveti, H., Maringanti, S., Parsa, A.: Polycystic ovary syndrome monitoring using machine learning. In 2023 International Conference on Intelligent Data Communication (2023)
28. Deshpande, S.S., Wakankar, A.: Automated detection of polycystic ovarian syndrome using follicle recognition. In: 2014 IEEE International Conference on Advanced Communications, Control and Computing Technologies, pp. 1341–1346. IEEE (2014)
29. Khanna, V.V., Chadaga, K., Sampathila, N., Prabhu, S., Bhandage, V., Hegde, G.K.: A distinctive explainable machine learning framework for detection of polycystic ovary syndrome. Appl. Syst. Innov. **6**(2), 32 (2023)
30. Dutta, P., Paul, S., Majumder, M.: An efficient SMOTE based machine learning classification for prediction & detection of PCOS (2021)
31. Maheswari, K., Baranidharan, T., Karthik, S., Sumathi, T.: Modelling of F3I based feature selection approach for PCOS classification and prediction. J. Ambient. Intell. Humaniz. Comput. **12**, 1349–1362 (2021)
32. Prapty, A.S., Shitu, T.T.: An efficient decision tree establishment and performance analysis with different machine learning approaches on polycystic ovary syndrome. In: 2020 23rd International Conference on Computer and Information Technology (ICCIT), pp. 1–5. IEEE (2020)
33. Hdaib, D., Almajali, N., Alquran, H., Mustafa, W.A., Al-Azzawi, W., Alkhayyat, A.: Detection of polycystic ovary syndrome (PCOS) using machine learning algorithms. In: 2022 5th International Conference on Engineering Technology and its Applications (IICETA), pp. 532–536. IEEE (2022)

34. Alagarsamy, M., Shanmugam, N., Mani, D.P., Thayumanavan, M., Sundari, K.K., Suriyan, K.: Detection of polycystic syndrome in ovary using machine learning algorithm. Int. J. Intell. Syst. Appl. Eng. **11**(1), 246–253 (2023)
35. Inan, M.S.K., Ulfath, R.E., Alam, F.I., Bappee, F.K., Hasan, R.: Improved sampling and feature selection to support extreme gradient boosting for PCOS diagnosis. In: 2021 IEEE 11th annual Computing and Communication Workshop and Conference (CCWC), pp. 1046–1050. IEEE (2021)
36. Meena, K., Manimekalai, M., Rethinavalli, S.: Correlation of artificial neural network classification and NFRS attribute filtering algorithm for PCOS data. Int. J. Res. Eng. Technol. **4**(3), 519–524 (2015)
37. Nasim, S., Almutairi, M.S., Munir, K., Raza, A., Younas, F.: A novel approach for polycystic ovary syndrome prediction using machine learning in bioinformatics. IEEE Access **10**, 97610–97624 (2022)
38. Denny, A., Raj, A., Ashok, A., Ram, C.M., George, R.: i-hope: detection and prediction system for polycystic ovary syndrome (PCOS) using machine learning techniques. In: TENCON 2019–2019 IEEE Region 10 Conference (TENCON), pp 673–678. IEEE (2019)
39. Rachana, B., Priyanka, T., Sahana, K.N., Supritha, T.R., Parameshachari, B.D., Sunitha, R.: Detection of polycystic ovarian syndrome using follicle recognition technique. Glob. Trans. Proc. **2**(2), 304–308 (2021)
40. Neto, C., Silva, M., Fernandes, M., Ferreira, D., Machado, J.: Prediction models for Polycystic Ovary Syndrome using data mining. In: International Conference on Advances in Digital Science, pp. 210–221. Springer, Cham (2021)
41. Munjal, A., Khandia, R., Gautam, B.: A machine learning approach for selection of polycystic ovarian syndrome (PCOS) attributes and comparing different classifier performance with the help of weka and pycaret. Int. J. Sci. Res. **9**, 1–5 (2020)
42. Zhang, X., Liang, B., Zhang, J., Hao, X., Xu, X., Chang, H.M., Leung, P.C., Tan, J.: Raman spectroscopy of follicular fluid and plasma with machine-learning algorithms for polycystic ovary syndrome screening. Mol. Cell. Endocrinol. **523**, 111139 (2021)
43. Soni, P., Vashisht, S.: Region of interest and bag of features for predicting polycystic ovaries in females. Think India J. **22**(3), 7901–7911 (2019)
44. Pushkarini, H., Anusuya, M.A.: A prediction model for evaluating the risk of developing PCOS. Int. Res. J. Eng. Technol. (IRJET) **7**(9), 1150–1156 (2020)
45. Thomas, N., Kavitha, A.: Prediction of polycystic ovarian syndrome with clinical dataset using a novel hybrid data mining classification technique. Int. J. Adv. Res. Eng. Technol. (IJARET) **11**(11), 1872–1881 (2020)
46. Tanwani, N.: Detecting PCOS using machine learning. Int. J. Mod. Trends Eng. Sci. (IJMTES) **7**(01), 2348–3121 (2020)
47. Raut, K., Katkar, C., Itkar, S.A.: PCOS detect using machine learning algorithm. Int. Res. J. Eng. Technol. **9**(01) (2022)
48. Vikas, B., Radhika, Y., Vineesha, K.: Detection of polycystic ovarian syndrome using convolutional neural networks. Int. J. Cur. Res. Rev. **13**(06), 156 (2021)
49. Kumari, S.: Classification of PCOS/PCOD using transfer learning and GAN architectures to generate pseudo ultrasound images. Doctoral Wisesty, U.N., 2016. Implementasi Gabor (2021)
50. Tiwari, S., Kane, L., Koundal, D., Jain, A., Alhudhaif, A., Polat, K., Zaguia, A., Alenezi, F., Althubiti, S.A.: SPOSDS: A smart polycystic ovary Syndrome diagnostic system using machine learning. Expert Syst. Appl. **203**, 117592 (2022)
51. Lakshmi, M.J., Spandana, D.S., Raj, H., Niharika, G., Kodipalli, A., Kamal, S., Rao, T.: Prediction of PCOS and PCOD in women using ML algorithms. In: International Conference on Information and Communication Technology for Intelligent Systems, pp. 97–115. Springer, Singapore (2023)
52. Vora, K., Shah, A., Shah, N., Verma, P.: Prediction of polycystic ovary syndrome (PCOS) using optimized machine learning classifiers. In: Proceedings of 3rd International Conference on Artificial Intelligence: Advances and Applications: ICAIAA 2022, pp. 1–12. Springer, Singapore (2023)

53. Saha, A., Roy, A., Chakraborty, B., Saha, B., Chowdhury, D.: A comparative study to predict polycystic ovarian syndrome (PCOS) based on different models of machine learning technique. Am. J. Electron. Commun. **4**(2), 1–6 (2023)
54. Kaur, N., Gupta, G., Hafiz, A., Sharma, M., Singh, J.: Machine learning algorithms for polycystic ovary syndrome/polycystic ovarian syndrome detection: a comparison. In: International Conference on Computer Vision and Robotics, pp. 453–465. Springer, Singapore (2023)
55. Sinthia, G., Poovizhi, T., Khilar, R.: Analysis on polycystic ovarian syndrome and comparative study of different machine learning algorithms. In: Advances in Intelligent Computing and Communication: Proceedings of ICAC 2021, pp. 191–196. Springer, Singapore (2022)
56. Kumar, K.S., Nirmala, V., Venkatalakshmi, K., Karthikeyan, K.: Analysis of optimization algorithms on follicles segmentation to support polycystic ovarian syndrome detection. J. Comput. Theor. Nanosci. **15**(1), 380–391 (2018)
57. Swapnarekha, H., Dash, P.B., Nayak, J., Routray, A.R.: An optimistic Bayesian optimization based extreme learning machine for polycystic ovary syndrome diagnosis. In: Nature-Inspired Optimization Methodologies in Biomedical and Healthcare, pp. 175–193. Springer, Cham (2022)
58. Prajna, K.B., Iyer, B.V., Bhuvan, C., Thambanda, K.M., Kanasu, H.R.: Implementation of various machine learning algorithms to predict polycystic ovary syndrome. In: 2023 4th International Conference for Emerging Technology (INCET), pp. 1–6. IEEE (2023)
59. Swamy, S.R., Nandini Prasad, K.S.: Hybrid machine learning model for early discovery and prediction of polycystic ovary syndrome. In: 2022 Second International Conference on Advanced Technologies in Intelligent Control, Environment, Computing & Communication Engineering (ICATIECE), pp. 1–8. IEEE (2022)
60. Rathod, Y., Komare, A., Ajgaonkar, R., Chindarkar, S., Nagare, G., Punjabi, N., Karpate, Y.: Predictive analysis of polycystic ovarian syndrome using CatBoost algorithm. In: 2022 IEEE Region 10 Symposium (TENSYMP), pp. 1–6 IEEE (2022)
61. Adla, Y.A.A., Raydan, D.G., Charaf, M.Z.J., Saad, R.A., Nasreddine, J., Diab, M.O.: Automated detection of polycystic ovary syndrome using machine learning techniques. In: 2021 Sixth international conference on advances in biomedical engineering (ICABME), pp. 208–212. IEEE (2021)
62. Katarya, R., Jindal, A., Duggal, A., Shah, A.: A novel polycystic ovarian syndrome diagnostic system using machine learning. In: Proceedings of 3rd International Conference on Computing Informatics and Networks: ICCIN 2020, pp. 555–563. Springer (2021)
63. Reka, S., Elakkiya, R.: Early diagnosis of poly cystic ovary syndrome (PCOS) in young women: a machine learning approach. In: 2022 IEEE International Symposium on Mixed and Augmented Reality Adjunct (ISMAR-Adjunct), pp. 286–288. IEEE (2022)
64. Marreiros, M., Ferreira, D., Neto, C., Witarsyah, D., Machado, J.: Classification of polycystic ovary syndrome based on correlation weight using machine learning. In: Big Data Analytics and Artificial Intelligence in the Healthcare Industry, pp. 150–176. IGI Global (2022)
65. Ahmetašević, A., Aličelebić, L., Bajrić, B., Bečić, E., Smajović, A., Deumić, A.: Using Artificial neural network in diagnosis of polycystic ovary syndrome. In: 2022 11th Mediterranean Conference on Embedded Computing (MECO), pp. 1–4. IEEE (2022)
66. Prasanna Kumar, H., Srinivasan, S.: Segmentation of polycystic ovary in ultrasound images. In: 2nd International Conference on Current Trends in Engineering and Technology, ICCTET, vol. 8, pp. 237–240 (2014)
67. Setiawati, E., Adiwijaya, Tjokorda, A.B.W.: Particle swarm optimization on follicles segmentation to support PCOS detection. In: 3rd International Conference on Information and Communication Technology (ICoICT), pp. 369–374 (2015)
68. Rihana, S., Moussallem, H., Skaf, C.: Automated algorithm for ovarian cysts detection in ultrasonogram. In: 2nd International Conference on Advances in Biomedical Engineering, pp. 219–222 (2013)
69. Sitheswaran, R., Dr. Malarkhodi, S.: An effective automated system in follicle identification for polycystic ovary syndrome using ultrasound images. In: International Conference on Electronics and Communication System (ICECS) (2014)

70. Raj, A.: Ovarian follicle detection for polycystic ovary syndrome using fuzzy C-means clustering. Int. J. Comput. Trends Technol. (IJCTT) **4**(7), 2146–2149 (2013)
71. Sterling. E.: Hormone levels and PCOS. Contemporary OB/GYN. (2011). https://www.contemporaryobgyn.net/view/hormone-levels-and-pcos. Accessed 16 May 2022
72. Kahsar-Miller, M.D., Azziz, R.: The effectiveness of the interview for predicting the presence of polycystic ovary syndrome. Gynecol. Endocrinol. **17**(6), 449–454 (2003). Sciencedirect.com. https://www.sciencedirect.com/science/article/abs/pii/S1751721415001396I. Accessed 16 May 2022
73. El Naqa, I., Murphy, M.J.: What is machine learning? In: Machine Learning in Radiation Oncology, pp. 3–11. Springer, Cham, Switzerland (2015)
74. Sidana, M.: Types of Classification Algorithms in Machine Learning. Medium Co.
75. Jakkula, V.: Tutorial on support vector machine (SVM), vol. 37, issue 2.5, p. 3. School of EECS, Washington State University (2006)
76. Nandipati, S.C., Ying, C.X., Wah, K.K.: Polycystic ovarian syndrome (PCOS) classification and feature selection by machine learning techniques. Appl. Math. Comput. Intell **9**, 65–74 (2020)
77. Rish, I., Watson, T.J., Center, R.: An empirical study of the Naive Bayes classifier. Accessed 16 May (2022)
78. Guo, G., Wang, H., Bell, D., Bi, Y., Greer, K.: KNN model-based approach in classification. In: On the Move to Meaningful Internet Systems 2003: CoopIS, DOA, and ODBASE: OTM Confederated International Conferences, CoopIS, DOA, and ODBASE 2003, Catania, Sicily, Italy, 3–7 Nov 2003. Proceedings, pp. 986–996. Springer (2003)
79. Ben-Haim, Y., Tom-Tov, E.: A streaming parallel decision tree algorithm. J. Mach. Learn. Re. **11**(2) (2010)
80. Aggarwal, S., Pandey, K.: Early identification of PCOS with commonly known diseases: obesity, diabetes, high blood pressure and heart disease using machine learning techniques. Expert Syst. Appl. **217**, 119532 (2023)
81. Biau, G., Scornet, E.: A random forest guided tour. Test **25**, 197–227. K-Means Clustering Algorithm (2016). www.javatpoint.com and https://www.javatpoint.com/k-meansclustering-algorithm-in-machine-learning. Accessed 16 May 2022
82. Wu, Y.C., Feng, J.W.: Development and application of artificial neural network. Wirel. Pers. Commun. **102**, 1645–2165 (2018)
83. Albawi, S., Mohammed, T.A., Al-Zawi, S.: Understanding of a convolutional neural network. In: 2017 International Conference on Engineering and Technology (ICET), pp. 1–6. IEEE (2017)
84. You, W., Shen, C., Wang, D., Chen, L., Jiang, X., Zhu, Z.: An intelligent deep feature learning method with improved activation functions for machine fault diagnosis. IEEE Access **8**, 1975–1985 (2019)
85. Bento, C.: Multilayer perceptron explained with a real-life example and python code: sentiment analysis. Towards Data Science (2021)

Mehak Abbas holds a Bachelor of Technology (B.Tech) degree in Electronics and Communication Engineering from Baba Ghulam Shah Badshah University, Rajouri, awarded in 2022. She further pursued a Master's degree in Electronics and Communication Engineering with a specialization in Digital Communication at Sharda University, Greater Noida, completing it in 2024. Her research focuses on the applications of Machine Learning and Deep Learning in healthcare and medical systems.

Shaheen Naz serves as an Assistant Professor in the Department of Electrical, Electronics, and Communication Engineering at the School of Engineering and Technology, Sharda University. With over 19 years of academic and research experience, her expertise lies in Material Science,

Wireless Communication, and Sensor Networks. Dr. Naz has an impressive publication record, with over 12 papers in renowned journals and three patents to her name.

Artificial Intelligence: A New Paradigm in Fetal Medicine Research and Clinical Practice

R. Sharan Kumar, K. Raghesh Krishnan, Neha Garg, and Reema Bhatt

Abstract This chapter examines the evolving role of Artificial Intelligence (AI) in antenatal fetal ultrasound, where it is increasingly integrated to enhance traditional ultrasound methods. AI applications, including Deep Learning (DL), Machine Learning (ML), and Convolutional Neural Networks (CNN), have proven effective in automating the acquisition of standard fetal ultrasound planes, refining biometric measurements, and aiding in the diagnosis of fetal abnormalities. These advancements optimize ultrasound procedures, reducing the time required for standard imaging and improving the accuracy, reliability, and efficiency of conventional approaches. This chapter explores the practical benefits, inherent challenges, and future potential of AI integration in fetal ultrasound, offering insights into its transformative impact on prenatal care.

Keywords Artificial intelligence (AI) · Fetal ultrasound · Deep learning (DL) · Machine learning (ML) · Convolutional neural networks (CNN) · Biometric measurements · Prenatal diagnostics · Healthcare information systems · Medical imaging accuracy · Data privacy in ultrasound

1 Introduction

Ultrasound has several applications in the prenatal context and is now considered the standard of care for all expectant mothers. It is increasingly being used for both regular monitoring and complicated pregnancy situations. Antenatal ultrasonography is non-invasive, non-radiative, incredibly convenient, and provides real-time imaging

R. Sharan Kumar · K. Raghesh Krishnan
Department of Computer Science and Engineering, Amrita School of Computing, Amrita Vishwa Vidyapeetham, Coimbatore, India

N. Garg · R. Bhatt (✉)
Department of Fetal Medicine and Perinatology, Amrita Institute of Medical Sciences and Research Centre, Faridabad, Haryana, India
e-mail: reemabhatt@fbd.amrita.edu

N. R. Vajjhala et al. (eds.), *Artificial Intelligence in Healthcare Information Systems—Security and Privacy Challenges*, Information Systems Engineering and Management 34, https://doi.org/10.1007/978-3-031-84404-1_4

[1–3]. It has been useful in assessing the growth of the fetus, identifying congenital abnormalities, and helping diagnosticians' image the fetus and its various parts. However, there are some limitations in using standard ultrasonography in clinical settings, including significant fetal movement, excessive maternal abdominal wall thickness, and inter-observer variability [4]. It is time-consuming and laborious for sonographers to perform precise measurements, routine illness diagnosis, and get high-quality standard biometric planes in accordance with standard protocols.

In recent years, Artificial Intelligence has been gradually applied in the medical field and in the field of fetal ultrasound, in particular [5]. The development of Artificial Intelligence (AI)-assisted ultrasound imaging is anticipated to accelerate the antenatal ultrasound examination [6, 7]. As a result, optimizing the prenatal ultrasound examination procedure reduces sonographers' workload, enhances clinical effectiveness, consistency, and shortens the examination time. AI has been effectively used to support traditional imaging techniques in the areas of automated fetal ultrasound standard plane recognition, biometric parameter measurement, and illness diagnosis. When it comes to accurately replicating values in ultrasound measurements and other repetitive ultrasound operations, AI shows great promise. Intelligent imaging may be used in many sophisticated ways with the most recent ultrasonic imaging technology. Additionally, it lessens the number of training cycles that new sonographers must complete, enhances clinical workflow quality control, encourages the prudent use of available medical resources, and reduces healthcare costs [7].

The use of the extensive data capacity of AI storage can be beneficial in identifying the risk factors associated with complicated pregnancies through the analysis of multiomics and large genomic data. The worldwide expansion of AI in healthcare systems has grown dramatically, driven by substantial investments in the field. Statistics indicate that investment in AI is predicted to increase from \$2.1 billion to \$36.1 billion by 2025 [8]. For successful inclusion of AI in healthcare, training and collaboration between partners are required. With one of the leading causes of death being medical errors, AI will help clinicians to make more confident and accurate decisions. However, it is not a substitute for clinical experience. Three-dimensional (3D) printing technology can create images that closely resemble real tissues; trainees may practice on a realistic model. It provides superior depth perception compared to two-dimensional (2D) models, enabling medical professionals to evaluate embryonic organs more accurately based on tissue depth and dimension.

2 What is Artificial Intelligence (AI)?

AI is the use of computer algorithms that mimic human intellect to solve issues or learn new information. The organization of AI is like the way that neurons are arranged in the human brain. These are known as artificial neural networks, comprising a reliable mathematical framework capable of analyzing complex, multifactorial data [9, 10]. A subset of Artificial Intelligence called Machine Learning

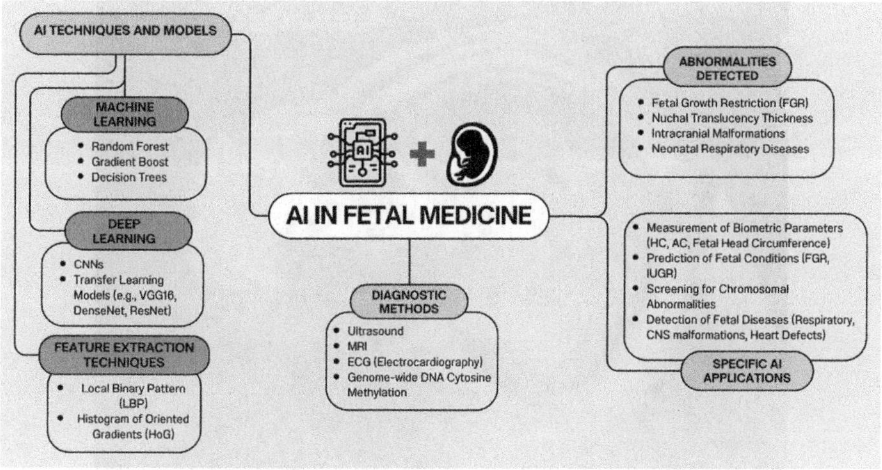

Fig. 1 Outline of AI technologies and their applications in fetal medicine

(ML) concentrates on learning and deriving rules from gathered data and drawing conclusions and making inferences about new data [11].

Deep Learning (DL) is a primary branch of ML that is significantly effective at target detection, image segmentation, and image classification. This contrasts with other forms of ML, which rely heavily on data quality and expertise. It has recently gained traction in medical imaging [12]. An overview of AI technologies and their uses in fetal medicine is shown in Fig. 1. DL algorithms utilize a convolutional neural network (CNN) with multiple hidden layers to extract essential features from a small number of training samples, resulting in high-performance predictions [13, 14].

3 AI Applications in the Measurement of Fetal Ultrasonic Biometry Parameters

3.1 Intelligent Fetal Head Circumference (HC) and Abdominal Circumference (AC) Measurement

For determining prenatal anomalies, analyzing fetal development, and determining gestational age and weight, HC and AC are crucial biometric indicators [15, 16]. The gestational age may be accurately estimated using these biometric measures and a variety of algorithms. Figure 2 illustrates Binder facies, a rare congenital disorder resulting in an underdeveloped face. Fetal morbidity and death can be decreased by increasing measurement precision. The accuracy of manual AC measurement can be impacted by variations in fetal position, oligohydramnios, and the thickness of the abdominal wall in pregnant women [7]. To lessen the workload on sonographers,

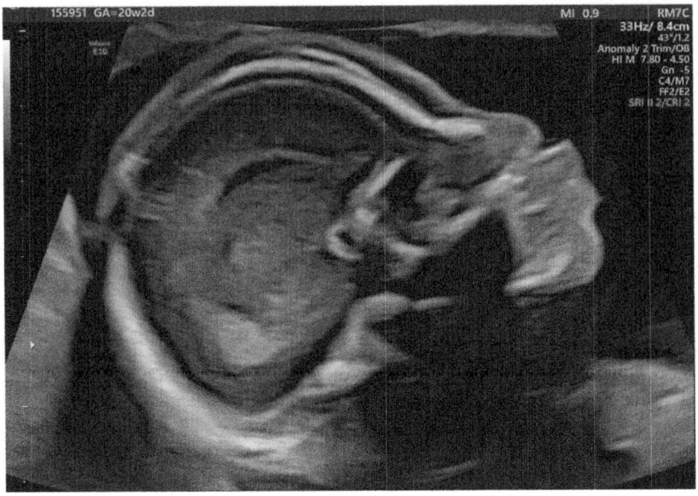

Fig. 2 Binder facies, a rare congenital disorder resulting in an underdeveloped face

a quick and precise way to measure AC is essential. Significant benefits have been demonstrated using convolutional neural networks in the categorization of medical images.

The research by Gornale et al. in [17] explores ultrasound image processing during the second and third trimesters of pregnancy in a novel way, utilizing cutting-edge deep learning techniques for fetal biometric analysis. The dataset archive consisting of 563 s trimester images and 750 third trimester images is used to estimate the head circumference (HC), abdominal circumference (AC), and femur length (FL), which are considered important measures for biometric analysis. A range of transfer learning classification techniques, including CNN, DNN (Deep Neural Network), ResNet50, LenNet, InceptionV2, XceptionNet, DenseNet121, MobileNet, MobileNetV2, and VGG16, were used to classify the segmented photos based on trimester parameters. The second trimester segmentation accuracies were reported as 69.09%, 92.02%, and 69.43% for AC, FL, and HC, respectively. In the third trimester, FL and HC attained 90.04% and 69.76% accuracy, respectively. MobileNet and XceptionNet were claimed to achieve remarkable accuracies of 99.8% across trimesters, demonstrating the robustness of the model, which indicates how Deep Learning might transform prenatal care by providing accurate image analysis.

Zeidan et al. studied fetal growth restriction (FGR) and investigated sophisticated MRI techniques that go beyond standard procedures to evaluate fetal health [18]. They postulated that knowledge gained from fetal liver and placental MRI would provide indicators essential for comprehending fetal growth restriction (FGR), a disorder associated with serious fetal problems. 12 FGR and 12 typically developed pregnancies were examined using diffusion-weighted MRI and T2 relaxometry. Multi-compartment models and Intravoxel Incoherent Motion (IVIM) were used to highlight the variations in the perfusion and oxygenation of the fetal organs. Important discoveries revealed unique characteristics in the placenta and fetal liver, indicating modified patterns of oxygenation and circulatory dynamics in FGR instances, which may have an impact on the supply of brain tissue to the fetus. Simple Machine Learning models analyzed voxel-level data for regression tasks, whereas a convolutional neural network diagnosed FGR and predicted delivery outcomes with 100% accuracy. This work appears to be a promising step towards the betterment of fetal health monitoring and intervention techniques by highlighting the potential of MRI-based biomarkers for improved FGR evaluation and diagnostic automation.

Using ML approaches for predictive modeling, novel research from a tertiary hospital in Bandar Abbas, Iran, from January 2020 to January 2022, used an inventive approach to address intrauterine growth restriction (IUGR), a major obstetric issue [19]. This cross-sectional study examined singleton pregnancies that lasted longer than 24 weeks, excluding multiple pregnancies and fetal abnormalities. Expert opinion and cohort studies were utilized to identify predictors, such as the mother's health and pregnancy-related characteristics. In order to predict the presence of IUGR, the study used Decision Tree Classification, Random Forest Classification, Deep Learning (DL), and Gradient Boost Algorithm. Models were evaluated based on Area Under the Receiver Operating Characteristic (AUROC), accuracy, precision, and sensitivity. The DL model showed an impressive AUROC of 0.91 and a 95% confidence interval (CI) ranging from 0.85 to 0.97, which was found to be the best among all that was tested. Prognostic variables included drug addiction, history of past IUGR, chronic hypertension, preeclampsia, maternal anemia, and the COVID-19 effect. This study showcases the promise of ML in obstetrics and stresses the need for further research to enhance IUGR diagnosis and treatment.

3.2 Intelligent Measurement of Fetal Nuchal Translucency (NT) Thickness

The subcutaneous fluid-filled region of the posterior fetal neck is known as nuchal translucency (NT) [7]. Chromosome abnormalities, infections, skeletal dysplasia, or developing heart diseases are thought to be linked to NT thickening during pregnancy. It is measured in the fetus's standard sagittal plane in accordance with criteria for accurate assessment, enabling the early diagnosis of genetic problems and fetal structural anomalies. However, the poor signal-to-noise ratio of ultrasound pictures, the length of the short fetal parietal rump, and the mobility of the fetus in the early stages of gestation provide difficulties for the usual plane collection of NT thickness. Compared to professionals, sonographers without experience devote 25.56% more time to critical biometric activities [20]. Experts from several fields have achieved significant advances in automated measuring in recent years.

Fetal abnormality detection for pregnancy management makes use of ultrasonography for screening of chromosomal abnormalities and genetic defects in the mid-trimester (18–22 weeks of gestation). Although most abnormalities could be diagnosed in the first trimester (11–14 weeks of gestation), detection rates are still found to be low due to the lack of experience of the radiographer. In [21], Computer Vision and Deep Learning models employ Nuchal Translucency (NT) as a marker. A test set consisting of 109 images with an NT thickness of 33 and a training set of 451 images with an NT thickness of 55 are used. The fetal and NT areas, or the ratio of NT to fetus, are segmented using an automated image segmentation approach using the U-Net CNN model. The methodology of this study consists of taking standardized mid-sagittal ultrasound images of the fetal head and chest from 560 fetuses between 11 and 13 weeks and 6 days of gestation. An NT thicker than 3.5 mm was present in 88 of them, or 15.7%. Two experts in fetal medicine graded the imaging quality as good or low. The segmentation efficacy is measured using Area Under the Receiver Operator Curve (AUROC). The test set segmentation resulted in a specificity of 91% and a dice score of 94% for thick NT instances. The detection rate in high-resolution images was found to be mathematically higher with $p = 0.003$ in comparison with low-resolution images.

Zhang et al. (2022) created and validated a DL model to reliably use ultrasound images for trisomy 21 (Down syndrome) screening [22]. Infants with Down syndrome have intellectual impairment and developmental delays. Since cell-free fetal DNA tests are expensive despite their excellent accuracy, a CNN model was created. In addition to Nuchal Thickness, nasal bone length and prenatal skin are used as markers. 2D ultrasound fetal faces acquired from 822 mothers with singleton pregnancy within 11–14 weeks of gestation and an average age of 31.9 years were split into a training set consisting of 550 samples and the validation set consisting of 272 samples and studied. Preprocessing in the form of Data segmentation and augmentation is carried out before input to the Deep Learning Model: Trisomy21Net, a shallow CNN model created to identify down syndrome. The screening efficiency is measured using ROC and AUC parameters, which demonstrated robust results in both the training (Area Under the Curve (AUC) of 0.98 and a 95% Confidence Interval (CI) ranging from 0.97 to 0.99).and validation (Area Under the Curve (AUC) of 0.95 and a 95% Confidence Interval (CI) ranging from 0.93 to 0.98.) sets. The NT marker and the mother's age were found to be the two factors contributing to the robust results of the approach.

To identify chromosomal abnormalities, such as Down syndrome (DS), S. Nirmala and V. Palanisamy focused their investigation on first-trimester fetal ultrasound images for assessing Nuchal Translucency (NT) thickness [23]. Increased fluid collection in the nuchal region (>2.5 mm) is a common symptom of trisomy. NT thickness of 13, 18, and 21 mm is found to be an important measure for screening fetuses with chromosomal abnormalities. Using ultrasound images acquired by an EMP 1100 machine, the research presents a semi-automatic computer-aided technique for NT measurement. Image preprocessing involves the segmentation of the NT region using Mean Shift and Canny operators. The topological thickness is finally assessed by blob analysis. This research not only provides a prospective solution for accurate NT quantification but also addresses the limitations of prior approaches such as edge detection and the need for fetal positioning restrictions.

4 AI Applications in Fetal Ultrasonic Diseases Diagnosis

4.1 AI Applications in Fetal Ultrasound of Neonatal Respiratory Diseases

While the research focusing on utilizing DL techniques to diagnose diseases in adults and children is extensive, the quantum of work focusing on neonates is scarce. Through the direct extraction of important characteristics from ultrasonic pictures and the efficient quantification of Fetal Lung Maturity FLM, texture feature analysis can lower examiner subjective variance. Respiratory disorders in neonates account for high infant mortality rates and demand speedy diagnosis and medical care. The work by Yildirim AE and Canayaz M in [24] attempts to fill this gap by proposing a novel Deep Learning-based approach for predicting respiratory disorders in neonates, especially focused on Neonatal ICU (Intensive Care Unit) patients using chest X-ray images. Neonatal X-ray pictures are classified using C + EffxNet, a hybrid Deep Learning model, to diagnose respiratory conditions such as pneumonia. Feature selection was performed using Principal Component Analysis followed by feature merging. The model's performance was measured using accuracy and kappa metrics. Prior to feature merging, the model reported an accuracy of 0.965 and a kappa value of 0.904, whereas after feature merging, the model achieved an accuracy of 0.977 and kappa value of 0.935. The model was also reported to perform remarkably well on a test dataset consisting of pediatric pneumonia cases, with an overall accuracy and kappa values of 0.992 and 0.982, respectively. According to experts, better treatment and survival during the first ten years of life are largely dependent on embracing this additional feature, and there should be more clinical trials on newborns, and the approach has a promising future.

A research study aimed at predicting respiratory support needs and mortality in extremely low birth weight preterm infants at 36 weeks postmenstrual age (PMA) employed ML techniques [25]. This significant study focused on bronchopulmonary dysplasia (BPD), a severe condition impacting the long-term health outcomes of preterm infants. Utilizing clinical data from the Taiwan Neonatal Network (TNN) database spanning from 2016 to 2019, the study included 3200 eligible infants based on specific birth weight (401–1500 g) and gestational age (22 weeks or 0 days to 31 weeks or 6 days) criteria. Application of logistic regression demonstrated the highest area under the receiver operating characteristic curve (AUROC) for predicting outcomes such as overall mortality, death before 36 weeks' PMA, and BPD severity in a new perspective. The study analyzed 24 early-life attributes and evaluated seven different Machine Learning algorithms such as k nearest neighbor, logistic regression, classification tree, random forest, naïve Bayes, neural network, and support vector machine. The model's AUROC values were found to be 0.801 for no BPD, 0.850 for severe BPD or death before 36 weeks of PMA, and 0.881 for total mortality. Although the study focused on the need for improving BPD definitions and predictions, it also demonstrated the suitability of Machine Learning models for clinical use, provided the attributes were chosen carefully.

The diagnosis of bronchopulmonary dysplasia (BPD), a chronic lung condition that affects both preterm and term newborns, is presently carried out by clinical observation and late-stage tests. In order to fill this gap, a novel approach for automatic lung segmentation in MRIs using Convolutional Neural Networks (CNNs) was developed, offering a radiation-free diagnostic substitute [26]. This study analyzed two distinct cohorts of preterm newborns, 48 male and 55 female, with a median gestational age of 26.57 weeks. 86 out of 103 children were diagnosed with BPD. The CNN models obtained a volumetric dice coefficient of 0.880 on an independent test set and 0.908 in cross-validation using quiet-breathing MRI data. The Lung volume, shape, pixel intensity, and surface generated from MRI scans showed strong relationships with lung damage ratings and were found to be useful in differentiating the severity of BPD levels (AUC 0.92 ± 0.02 for BPD vs. no BPD, AUC 0.84 ± 0.03 for no/mild BPD vs. moderate/severe BPD). The results highlight the performance and promise of Deep Learning models for non-invasive infant lung illness evaluation, despite the study's restriction to MRI (Magnetic resonance imaging) data.

4.2 AI Applications in Fetal Ultrasound of Intracranial Malformations

The study of the developing brain to identify and treat prenatal brain abnormalities is known as fetal neurosonography. The process involves physically capturing brain regions and photos. The final diagnosis can be impacted by a number of factors, including the sonographer's experience and skill, the thickness of the mother's abdominal wall, and the unfavorable fetal head position. High false-positive and false-negative rates may result from this. Improved fetal anomaly detection rates can result from overcoming these constraints with the use of AI-assisted models. Figure 3 depicts a comprehensive evaluation of the brain, including the ventricles, cortex, and other critical structures.

The remarkable work by Lin et al. in [27] resulted in a ground-breaking project called PAICS (Prenatal Artificial Intelligence for CNS Screening), an AI-powered system designed to improve the ultrasound imaging-based identification of fetal central nervous system (CNS) abnormalities. Using a large dataset, which included external and prospective validation datasets, as well as 4202 images from abnormal pregnancies and 1426 from normal pregnancies for internal validation, the researchers carefully preprocessed the images using methods such as flipping, rotating, and resizing to 1600×900 pixels. Utilizing a powerful NVidia GeForce GPU configuration, the work used the YOLO v3 algorithm, continuously training

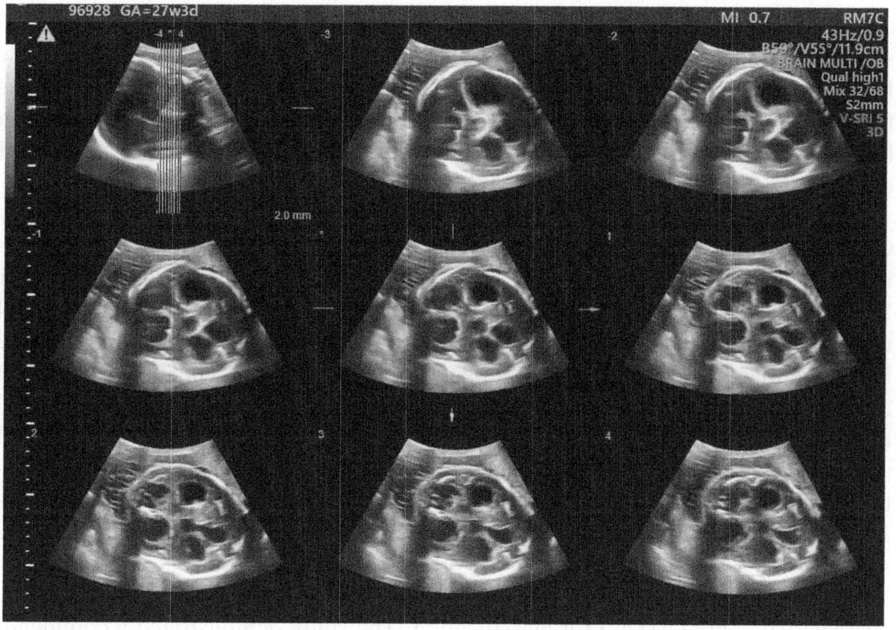

Fig. 3 Comprehensive evaluation of the brain, including the ventricles, cortex, and other critical structures

and optimizing it over a period of four days. When the model's performance was assessed using metrics such as accuracy, sensitivity, specificity, AUC, and Fleiss's kappa, it was found to have good diagnostic efficiency and accuracy. The study did identify certain shortcomings, though, such as streamlined training protocols, constrained sample sizes, and untested gestational age variances. Despite these limitations, PAICS appears to be a major advancement, promising improved prenatal diagnostic skills and providing sonologists with a reliable tool for second opinion.

In a study claimed to be the first of its kind, Lin et al. in [28] analyzed the performance of a Deep Learning (DL) system, PAICS (Prenatal Ultrasound Diagnosis Artificial Intelligence Conduct System) in detecting intracranial abnormalities in fetuses. A database of 709 neurosonographic images and videos from 558 fetuses was used to evaluate PAICS to differentiate nine types of fetal intracranial pathologies together with normal image patterns. A three-way, crossover randomized, controlled trial was followed where 36 sonologists with varying levels of experience reviewed the images/videos in three formats: unaided, concurrent (those randomized to view with PAICS at the beginning), and second (those who viewed with PAICS after initial unaided assessment). The study revealed that PAICS led to a significant increase in the diagnostic accuracy from 73% unassisted to 80% with concurrent and 82% with second mode. The AUC scores were found to improve, starting from 0.85 up to 0.89 and then 0.90. The investigation did not explore the specific architecture or training of PAICS, though. The placement of PAICS, however, affected reading time to some degree as well, with a small difference in concurrent mode and a huge difference in secondary mode. Despite these variations, the study discovered that PAICS can allow for a substantial increase in diagnostic performance, where the concurrent mode was found to be the most efficient. The study demonstrated the feasibility of using AI in healthcare, with its focus on the enhancement of accuracy and efficiency in prenatal diagnosis.

Xie et al. in [29] investigated the important scientific topic of how effectively CNNs could categorize fetal central nervous system abnormalities, a highly specialized domain, as normal or abnormal, using fetal brain ultrasonography images. Over 22,000 axial slices images from a hospital database consisting of 10,251 normal and 2529 abnormal pregnancy cases were carefully pre-processed and augmented into more than 28,000 standard axial planes of each category. The dataset was split at case level into a training and testing ratio of 8:2, emphasizing three essential tasks namely lesion region segmentation, abnormality classification (a binary label assignment indicating the presence or absence of abnormality) and lesion localization within the abnormal images. The study made use of Deep Learning algorithms in the form of CNNs, to evaluate the accuracy of automated lesion localization as measured by (i) expert annotation (ii) precision, recall and Dice's coefficient for lesion segmentation (iii) sensitivity, specificity and area under receiver operating characteristic curve (AUC) for classification. The study demonstrated a commendable performance, achieving an overall classification accuracy of 96.3%, a sensitivity of 96.9% and a specificity of 95.9%. The lesions were correctly localized in 61.6% of abnormal cases, closely localized in 24.6% cases, and irrelevant localization was observed in 13.7% cases. These findings offer a robust foundation for the development of more in-depth research aimed toward the diagnosis of fetal intracranial abnormalities using Deep Learning. The study provides hope that such approaches can potentially lead to better prenatal diagnosis as well as underscoring the power of Deep Learning in medical image analysis.

4.3 AI Applications in Fetal Ultrasound of Congenital Heart Diseases

The operator-dependent examination of the fetal heart used in fetal echocardiography is used to identify and classify congenital heart abnormalities. Figure 4 shows an ultrasound image of a cardiac rhabdomyoma, a noncancerous tumor caused by genetic changes. The left image (a) depicts the fetus, and the right image (b) is an annotated image of the Cardiac rhabdomyoma. Due to the fetus's constant mobility, its heartbeat occurring more quickly than that of an adult, its smaller heart than an adult's, and the lengthy learning curve for specialized knowledge, accurately identifying and localizing lesions is challenging and time-consuming [30, 31]. These

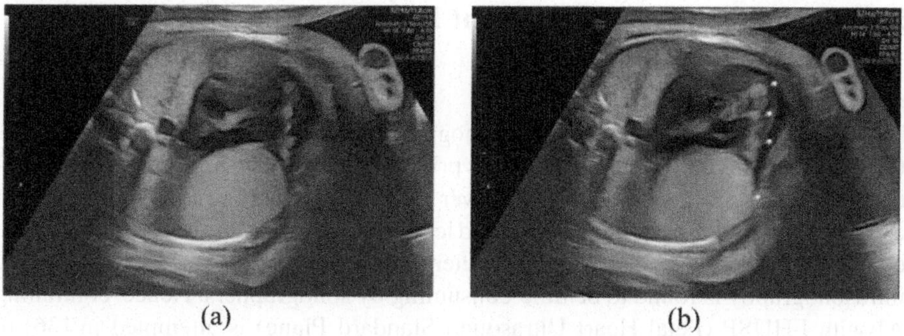

Fig. 4 Cardiac rhabdomyoma, a noncancerous tumor caused by genetic changes

shortcomings result in a high incidence of missed diagnoses, which postpones treatment and lowers prognosis. The issues should be resolved when AI and conventional ultrasonography are combined [32, 33].

An increase in the number of congenital cardiac illnesses demands the need for advanced techniques and efficient algorithms to enable earlier diagnosis and effective treatments. Vries et al. [34] couple AI and ECG to detect cardiac disorders in fetuses. A total of 122 patients—65 healthy individuals and 57 CHD cases—were taken into consideration for this investigation. To ensure effective algorithm performance, the Bayesian update rule was applied. The model's results turned out to be 71% for specificity, 63% for sensitivity, and 71% for accuracy. A noteworthy result was seen in the study when at least 14 Electrocardiogram segments (37.5 min) were used. Empirical evidence indicates a little negative correlation between gestational age and accuracy.

A study investigating the effectiveness of AI analysis using genome-wide DNA (Deoxyribonucleic Acid) cytosine methylation used the idea of cell-free DNA to diagnose CHD [35]. The study used twelve cases of distinct non-syndromic congenital cardiac abnormalities and twenty-six matched controls. The Illumina Infinium Methylation EPIC BeadChip array is a vital tool for preprocessing whole-genome cytosine nucleotide methylation studies. To uncover gene pathways that are crucial for CHD pathology, the ingenuity pathway analysis is also performed. With 98% sensitivity and 94% specificity, a total of 5 whole-genome cytosine nucleotide markers produced an AUROC value of 0.97. This work greatly contributed to the understanding of epigenetic changes in fetuses.

5 AI Applications in Intelligent Detection of the Fetal Ultrasonic Standard Plane

The proficiency and experience of the sonographers are crucial for the precise acquisition of fetal structural planes. This skill's primary drawback is the considerable intra-observer variability and poor inter-observer similarity of the obtained pictures. AI can assist in obtaining these planes ideally and lessen the stated restrictions with its feature extraction capabilities. For example, determining the fetal planes in fetal cardiac ultrasonography is found to be time-consuming by sonographers. Hence, combining AI with FHUSP (Fetal Heart Ultrasound Standard Plane) is attempted in [36] to improve the precision and effectiveness of diagnosis. In addition to 788 standard plane data sets, the collection also includes 448 normal planes and abnormal plane data sets. The model integrates textural characteristics of Local Binary Pattern (LBP) and Histogram of Oriented Gradients (HoG) with the Bag of Words (BOW) model for feature fusion, and classification is performed using Support Vector Machine (SVM). Direct correlation techniques are used in addition to LBP and HoG. The model's performance reported a classification accuracy of 87.35% and an average accuracy of 84.92% in the classification of normal and abnormal pla. In addition to pure AI-based techniques, AI coupled with image processing techniques show promising results in the field of medical image analysis [37, 38]. The incorporation of such practices in the fetal domain could result in enhancement of outcomes as shown in Table 1.

6 Challenges and Advancements

Despite the appreciable performance of AI technology in fetal medicine, there exist various challenges and opportunities for advancements.

6.1 Challenges in AI Integration

Data Accessibility and Data Quality

- **Small datasets**: Scarcity of high-quality, annotated fetal images with adequate heterogeneity for training the models.
- **Data Privacy**: Anonymity of patients and fetal health information in training, testing, and exchange of results is of prime importance.
- **Data Bias**: AI systems could have a bias towards certain communities and groups in a society if the data used does not reflect all the people or is inadequate.
- **Black-Box Models**: Some of the AI models, such as the Deep Learning models, are "black box models", and the process by which they arrive at a certain decision is hard to comprehend. A lack of decision transparency influences the clinician's trust to implement the respective AI technology.

Table 1 Summary of AI applications and accuracy in fetal biometry and diagnosis

Topic	Type of abnormality	Methods used (detailed)	Outcomes/accuracy	Sample size	Technological platform used	Challenges noted
Fetal Ultrasonic Biometry Parameters	Head Circumference (HC), Abdominal Circumference (AC), Femur Length (FL)	Convolutional Neural Networks (CNN), Deep Neural Networks (DNN), ResNet50, LenNet, InceptionV2, XceptionNet, DenseNet121, MobileNet, VGG16	Accuracies: HC (69.43–69.76%), AC (92.02–69.09%), FL (90.04%)	1313 images	Various deep learning frameworks	Variability due to sonographer experience
Fetal Growth Restriction (FGR)	Fetal Liver and Placental MRI	MRI techniques, Convolutional Neural Networks (CNN) used for analysis of MRI images to diagnose FGR	Diagnosis and prediction with 100% accuracy in CNN model	24 pregnancies	MRI, AI analytics platforms	High complexity of MRI data interpretation
Nuchal Translucency (NT) Thickness	Chromosomal abnormalities	U-Net Convolutional Neural Network model for segmentation, Trisomy21Net for Down syndrome screening	Specificity of 91%, AUC of 0.95–0.98	822 mothers	2D Ultrasound, CNN models	Difficulty in early gestation detection

(continued)

Table 1 (continued)

Topic	Type of abnormality	Methods used (detailed)	Outcomes/accuracy	Sample size	Technological platform used	Challenges noted
Neonatal Respiratory Diseases	Pneumonia in Neonates	Hybrid deep learning model (C + EffxNet) using Principal Component Analysis (PCA) and feature merging	Accuracy up to 0.992, Kappa up to 0.982	Pediatric case dataset	X-ray imaging, AI analysis software	Handling diverse clinical presentations
Fetal Ultrasonic Diseases	Intracranial malformations	PAICS (Prenatal Artificial Intelligence for CNS Screening), deep learning systems used to enhance imaging-based identification of abnormalities	Diagnostic accuracy up to 82%, AUC up to 0.90	Hundreds of images and videos	High-performance computing, deep learning models	False positives and negatives
Congenital Heart Diseases (CHD)	Congenital heart defects	AI integrated with electrocardiography (ECG), genome-wide DNA cytosine methylation analysis for diagnosing CHD	Sensitivity up to 98%, Specificity up to 94%, AUROC 0.97	148 patients	ECG, genomic analysis platforms	Accuracy vs. gestational age

- **Clinical Validation**: The above point makes it crucial that the AI models go through formal and predefined clinical validations to have accurate outcomes. This becomes costly and cumbersome at times, as positive research outcomes and effective therapeutic assets are mutually exclusive.

Integrating Clinical Workflow with Integration

- **Opposition to Adoption**: The primary reasons for some physicians' hesitancy to adopt AI tools in clinical practice may be due to the lack of AI knowledge, reluctance to change in workstyle, and fear of replacement.
- **Disruption of Workflow**: Implementation of AI in clinical practice might involve risks that upset the locally instituted practices and protocols. Hence, AI tools must complement the existing procedures without complicating or overburdening the prevailing therapeutic processes.
- **Education and Training**: Imparting required knowledge and skills to medical professionals to make them efficient in using the AI-related technologies becomes essential to decrease the gap between both domains.

6.2 Advancements in AI Integration

- Establishing international consortia to enhance collaboration on standardization, collection, and sharing of high-quality, legalized fetal medicine datasets. This will help overcome the scarcity of diverse and substantial datasets.
- Utilizing cutting-edge technology for anonymization and cryptography to protect data privacy while preserving its usefulness for AI research and training, thereby ensuring datasets are legally compliant.
- Developing and deploying advanced algorithms intended to detect and address biases in AI training data, thus ensuring fair AI results for various demographics.
- Investing in the development and deployment of XAI (Explainable AI) techniques to make the decision-making processes of AI models transparent and building trust among physicians and patients.
- Lowering the cost and complexity involved in bringing AI solutions to market by establishing streamlined regulatory routes for faster validation and certification of AI technology in healthcare.
- Developing simulation software or tools that allow healthcare providers to interact with AI-monitored systems in a controlled environment, enhancing comfort and understanding of these technologies.

7 Conclusion

Advances in AI and ML are revolutionizing the landscape of neonatal care by aiding in vital sign monitoring, prediction of health conditions, neurological function support, and infection detection using image analysis. This study highlights the impact of incorporating AI technologies with medical diagnosis. The examined literature throws light on the role of AI in revolutionizing neonatal medicine, especially in retinopathy of prematurity (ROP) and general newborn care. ML, DL, and CNNs are some of the techniques used in AI applications, which also include enhanced image identification, neurological diagnostics, and disease prediction. Standardized performance assessments using performance metrics such as sensitivity, accuracy, ROC, AUROC, etc. and high-quality data in adequate volume prove to be essential for successful AI integration. The study promises that AI has the potential to improve ROP screening through telemedicine, particularly in places with low resources. However, barriers such as the requirement for validation, digital literacy of the healthcare personnel, and encouraging teamwork require surmounting. To ensure safety and dependability in clinical practice, challenges such as dataset size, generalizability, and ethical issues such as data privacy and algorithm transparency call for sustained study and validation. Even though the studies offer incremental improvements in results, they show promising progress with significant therapeutic implications. Development of such models and algorithms would ultimately reduce the risk and dependency on expert sonographers, resulting in consistent fetal health monitoring systems, efficient diagnosis, and overall prenatal care, contributing to intelligent and life-saving medical healthcare systems.

References

1. Yang, X., Wang, X., Wang, Y., Dou, H., Li, S., Wen, H., Lin, Y., Heng, P.-A., Ni, D.: Hybrid attention for automatic segmentation of whole fetal head in prenatal ultrasound volumes. Comput. Methods Programs Biomed. **194**, 105519 (2020)
2. Ghelich Oghli, M., Shabanzadeh, A., Moradi, S., Sirjani, N., Gerami, R., Ghaderi, P., Sanei Taheri, M., Shiri, I., Arabi, H., Zaidi, H.: Automatic fetal biometry prediction using a novel deep convolutional network architecture. Phys. Med. **88**, 127–137 (2021)
3. Akkus, Z., Cai, J., Boonrod, A., Zeinoddini, A., Weston, A.D., Philbrick, K.A., Erickson, B.J.: A survey of deep-learning applications in ultrasound: artificial intelligence-powered ultrasound for improving clinical workflow. J. Am. Coll. Radiol. **16**, 1318–1328 (2019)
4. Fiorentino, M.C., Villani, F.P., Di Cosmo, M., Frontoni, E., Moccia, S.: A review on deep-learning algorithms for fetal ultrasoundimage analysis. Med. Image Anal. **83**, 102629 (2023)
5. Chen, Z., Liu, Z., Du, M., Wang, Z.: Artificial Intelligence in obstetric ultrasound: an update and future applications. Front. Med. **8**, 733468 (2021)
6. Dawood, Y., Buijtendijk, M.F., Shah, H., Smit, J.A., Jacobs, K., Hagoort, J., Oostra, R.-J., Bourne, T., Hoff, M.J.V.D., de Bakker, B.S.: Imaging fetal anatomy. Semin. Cell Dev. Biol. **131**, 78–92 (2022)
7. He, F., Wang, Y., Xiu, Y., Zhang, Y., Chen, L.: Artificial intelligence in prenatal ultrasound diagnosis. Front. Med. **8**, 729978 (2021)

8. Iftikhar, P., Kuijpers, M.V., Khayyat, A., Iftikhar, A., De Sa, M.D.: Artificial intelligence: a new paradigm in obstetrics and gynecology research and clinical practice. Cureus **12**(2) (2020)
9. It's a No Brainer: An Introduction to Neural Networks. (2019). Accessed 26 Jan 2020
10. Desai, G.S.: Artificial intelligence: the future of obstetrics and gynecology. J. Obstetr. Gynecol. India **68**, 326–327 (2018)
11. Hamet, P., Tremblay, J.: Artificial intelligence in medicine. Metab. Clin. Exp. **69S**, S36–S40 (2017)
12. Litjens, G., Kooi, T., Bejnordi, B.E., Setio, A.A.A., Ciompi, F., Ghafoorian, M., van der Laak, J.A.W.M., van Ginneken, B., Sánchez, C.I.: A survey on deep learning in medical image analysis. Med. Image Anal. **42**, 60–88 (2017)
13. Drukker, L., Noble, J.A., Papageorghiou, A.T.: Introduction to artificial intelligence in ultrasound imaging in obstetrics and gynecology. Ultrasound Obstet. Gynecol. **56**, 498–505 (2020)
14. Garcia-Canadilla, P., Sanchez-Martinez, S., Crispi, F., Bijnens, B.: Machine learning in fetal cardiology: what to expect. Fetal Diagn. Ther. **47**, 363–372 (2020)
15. Li, J., Wang, Y., Lei, B., Cheng, J.-Z., Qin, J., Wang, T., Li, S., Ni, D.: Automatic fetal head circumference measurement in ultrasound using random forest and fast ellipse fitting. IEEE J. Biomed. Health Inform. **22**, 215–223 (2018)
16. Sobhaninia, Z., Rafiei, S., Emami, A., Karimi, N., Najarian, K., Samavi, S., Reza Soroushmehr, S.M.: Fetal ultrasound image segmentation for measuring biometric parameters using multi-task deep learning. In: Annual International Conference of the IEEE Engineering in Medicine and Biology Society 2019, pp. 6545–6548 (2019)
17. Gornale, S., Kamat, P., Siddalingappa, R., Kumar, S.: Deep learning techniques for a comprehensive analysis of fetal biometric parameters across trimesters (2024)
18. Mutaz Zeidan, A., Ramirez Gilliland, P., Patel, A., Ou, Z., Flouri, D., Mufti, N., Maksym, K., Aughwane, R., Ourselin, S., David, A., Melbourne, A.: PIPPI2021: An approach to automated diagnosis and texture analysis of the fetal liver & placenta in fetal growth restriction (2022). arXiv e-prints, arXiv-2211
19. Taeidi, E., Ranjbar, A., Montazeri, F., Mehrnoush, V., Darsareh, F.: Machine learning-based approach to predict intrauterine growth restriction. Cureus **15**(17) (2023)
20. Yasrab, R., Fu, Z., Zhao, H., Lee, L.H., Sharma, H., Drukker, L., Papageorgiou, A.T., Alison Noble, J.: A machine learning method for automated description and workflow analysis of first trimester ultrasound scans. IEEE Trans. Med. Imaging (2022)
21. Kasera, B., Shinar, S., Edke, P., Pruthi, V., Goldenberg, A., Erdman, L., Van Mieghem, T.: Deep-learning computer vision can identify increased nuchal translucency in the first trimester of pregnancy. Prenat. Diagn. **44**, 535–543 (2024)
22. Zhang, L., Dong, D., Sun, Y., Hu, C., Sun, C., Wu, Q., Tian, J.: Development and validation of a deep learning model to screen for trisomy 21 during the first trimester from nuchal ultrasonographic images. JAMA Netw. Open **5**(6), e2217854–e2217854 (2022)
23. Nirmala, S., Palanisamy, V.: Measurement of nuchal translucency thickness for detection of chromosomal abnormalities using first trimester ultrasound fetal images (2010). arXiv:1001.1986
24. Yildirim, A.E., Canayaz, M.: A novel deep learning-based approach for prediction of neonatal respiratory disorders from chest X-ray images. Biocybern. Biomed. Eng. **43**(4), 635–655 (2023)
25. Wu, T.Y., Lin, W.T., Chen, Y.J., Chang, Y.S., Lin, C.H., Lin, Y.J.: Machine learning to predict late respiratory support in preterm infants: a retrospective cohort study. Sci. Rep. **13**(1), 2839 (2023)
26. Mairhörmann, B., Castelblanco, A., Häfner, F., Koliogiannis, V., Haist, L., Winter, D., Flemmer, A., Ehrhardt, H., Stöcklein, S., Dietrich, O. and Förster, K., Schubert, B.: Automated MRI lung segmentation and 3D morphologic features for quantification of neonatal lung disease. Radiol.: Artif. Intell. **5**(6), e220239 (2023)
27. Lin, M., He, X., Guo, H., He, M., Zhang, L., Xian, J., Lei, T., Xu, Q., Zheng, J., Feng, J. Hao, C., Xie, H.: Use of real-time artificial intelligence in detection of abnormal image patterns in standard sonographic reference planes in screening for fetal intracranial malformations. Ultrasound in Obstetr. Gynecol. **59**(3), 304–316 (2022)

28. Lin, M., Zhou, Q., Lei, T., Shang, N., Zheng, Q., He, X., Wang, N., Xie, H.: Deep learning system improved detection efficacy of fetal intracranial malformations in a randomized controlled trial. NPJ Digital Med. **6**(1), 191 (2023)
29. Xie, H.N., Wang, N., He, M., Zhang, L.H., Cai, H.M., Xian, J.B., Xian, Y.Z., Lin, M.F., Zheng, J., Yang, M.F.: Using deep-learning algorithms to classify fetal brain ultrasound images as normal or abnormal. Ultrasound Obstet. Gynecol. **56**(4), 579–587 (2020)
30. Raj, C.G., Harsha, V.S., Gowthami, B.S., Sunitha, R.: Virtual instrumentation based fetal ECG extraction. Procedia Comput. Sci. **70**, 289–295 (2015)
31. Karmegaraj, B., Kumar, S., Srimurugan, B., Sudhakar, A., Simpson, J.M., Vaidyanathan, B.: 3D/4D spatiotemporal image correlation (STIC) fetal echocardiography provides incremental benefit over 2D fetal echocardiography in predicting postnatal surgical approach in double-outlet right ventricle. Ultrasound Obstet. Gynecol. **57**(3), 423–430 (2021)
32. Bridge, C.P., Ioannou, C., Noble, J.A.: Automated annotation and quantitative description of ultrasound videos of the fetal heart. Med. Image Anal. **36**, 147–161 (2017)
33. Akkus, Z., Aly, Y.H., Attia, I.Z., Lopez-Jimenez, F., Arruda-Olson, A.M., Pellikka, P.A., Pislaru, S.V., Kane, G.C., Friedman, P.A., Oh, J.K.: Artificial intelligence (AI)-empowered echocardiography interpretation: a state-of-the-art review. J. Clin. Med. **10**, 1391 (2021)
34. de Vries, I., van Laar, J., van der Hout-van der Jagt, M., Clur, S., Vullings, R.: Fetal electrocardiography and artificial intelligence for prenatal detection of congenital heart disease. Acta obstetricia et gynecologica Scandinavica. **102**(11), 1511–1520 (2023)
35. Bahado-Singh, R,, Friedman, P., Talbot, C., Aydas, B., Southekal, S., Mishra, N., Guda, C., Yilmaz, A., Radhakrishna, U., Vishweswaraiah, S.: Cell-free DNA in maternal blood and artificial intelligence: accurate prenatal detection of fetal congenital heart defects. Am. J. Obstet. Gynecol. **228**(1), 76–e1.
36. Wu, B., Liu, P., Wu, H., Liu., S., He, S., Lv, G.: An effective machine-learning based feature extraction/recognition model for fetal heart defect detection from 2D ultrasonic imageries. Comput. Model. Eng. Sci. (CMES) **134**(2) (2023)
37. Midhila, M., Krishnan, K.R., Sudhakar, R.: A study of the phases of classification of liver diseases from ultrasound images and gray level difference weights based segmentation. In: 2017 International Conference on Communication and Signal Processing (ICCSP), pp. 1582–1587. IEEE (2017)
38. Raghesh Krishnan, K., Sudhakar, R.: Automatic classification of liver diseases from ultrasound images using GLRLM texture features. In: Soft Computing Applications: Proceedings of the 5th International Workshop Soft Computing Applications (SOFA), pp. 611–624. Springer (2013)

R. Sharan Kumar a researcher at Amrita Vishwa Vidyapeetham, Coimbatore, specializes in Computer Science and Engineering with a focus on the applications of Artificial Intelligence in medical imaging, particularly fetal medicine. Sharan's research delves into enhancing traditional ultrasonography through AI, aiming to expedite examination processes and improve the reliability of fetal assessments. His work represents an important step towards integrating advanced technologies in critical healthcare applications.

K. Raghesh Krishnan Assistant Professor at Amrita School of Engineering, Coimbatore, specializes in Image Processing and Biomedical Image Classification. His research focuses on advancing medical image analysis, with notable contributions in liver disease classification from ultrasound imaging. His work has received substantial citations in scientific literature, reflecting his significant impact in the field.

Neha Garg serves as a Consultant in Fetal Medicine at Moolchand Hospital, New Delhi. She holds an MS in Obstetrics and Gynecology and a Fellowship in Fetal Medicine from the Amrita Institute. Renowned for her dedication to patient care, Dr. Garg has published extensively in national and international journals and has received numerous accolades, including a recent silver medal for her research at the Society of Fetal Medicine (SFM) conference.

Reema Bhatt is the esteemed Head of Fetal Medicine and Perinatology at the Amrita Institute of Medical Sciences and Research Centre, Faridabad. With over 23 years of service in the Armed Forces, she has been honored with numerous prestigious awards, including the President's Appreciation. Dr. Bhatt is a prominent figure in fetal medicine, serving as the National Executive Secretary of the Society of Fetal Medicine and overseeing a successful fellowship program dedicated to advancing expertise in her field.

Trends in AI for Disease and Diagnostic Prediction: A Healthcare Perspective

Saurav Kumar Mishra, Tabsum Chhetri, Anagha Balakrishnan, Sneha Roy, Kusum Gurung, J. Jeba Praba, and John J. Georrge

Abstract This chapter explores the transformative impact of artificial intelligence (AI) on the healthcare system, particularly in enhancing the accuracy, efficiency, and speed of disease diagnostics. A key advantage of AI integration in healthcare lies in its capacity to analyze vast amounts of medical data with unprecedented precision, enabling early disease detection and supporting proactive interventions. By comparing diagnostic results with extensive datasets, AI algorithms can identify early-stage or dormant conditions, such as suspicious tumors or lesions, which facilitates early intervention and reduces the risk of disease progression. Despite these advancements, challenges persist regarding data security and patient privacy, as well as the ethical application of algorithms in clinical practice. Effective AI implementation also requires thorough training of healthcare professionals to interpret AI-driven insights accurately. Nonetheless, AI-driven healthcare holds substantial potential for advancing patient care and expanding medical understanding. This chapter provides a comprehensive analysis of recent AI advancements in disease prediction and diagnostics, highlighting its essential role in the future of healthcare.

Keywords Artificial intelligence (AI) · Healthcare diagnostics · Disease prediction · Early detection · Medical data analysis · Patient privacy · Ethical AI in healthcare · Proactive intervention · Data security · AI-driven patient care

---

S. K. Mishra · T. Chhetri · A. Balakrishnan · S. Roy · K. Gurung · J. J. Georrge (✉)
Department of Bioinformatics, University of North Bengal, Siliguri, India
e-mail: johngeorrge@gmail.com

J. Jeba Praba
Department of Computer Science and Applications, Christ College, Rajkot, India

N. R. Vajjhala et al. (eds.), *Artificial Intelligence in Healthcare Information Systems—Security and Privacy Challenges*, Information Systems Engineering and Management 34, https://doi.org/10.1007/978-3-031-84404-1_5

1 Introduction

Artificial Intelligence (AI) is the field of computer science that has impacted the world with positive changes. It focuses on developing machines and algorithms that imitate cognitive functions for real-world problem-solving and developing technologies that learn and think like people [1]. In recent years, the medical field has also had many advancements. The integration of AI has been the cornerstone for such advancements (Fig. 1). The main objective of the healthcare industry is disease diagnosis, and to diagnose the disease accurately, it is essential to interpret the symptoms and indications of an individual. An inaccurate diagnosis can lead to missed or postponed treatment, and in an unfortunate incident, an individual can lose their life. Therefore, the diagnosis is a complex and crucial process that requires an amalgamation of physical examination, medical history, and various diagnostic tests [2]. Diagnosing a disease can be complex, laborious, and time-consuming as specific medical experts are well-equipped with knowledge in their field but need to gain knowledge in another domain [3]. To address this, an automatic diagnostic system is required, combining computer accuracy and human understanding.

In this context, artificial neural networks (ANN) are one of the AI-based networks used in diagnostics as they function similarly to the human brain and use multilayered connections that link input and output to generate accurate results [4]. Subsequently, AI algorithms, such as Deep Learning (DL) and Machine Learning (ML), are now being used in the medical industry to identify and manage diseases [5]. AI systems can integrate data from different domains, handle bulk data, read the patterns that health professionals may miss, and predict the disease before its onset. AI can also improve hospital operations by anticipating needs, streamlining administrative work, and allocating resources optimally [6, 7]. Subsequently, AI can also tailor medicine according to individual needs. Analyzing patient data can also facilitate preventive and individualized care by identifying risk factors and modifying treatment strategies [8]. AI algorithms can examine patient data to generate customized care plans and treatment recommendations from Remote Patient Monitoring (RPM) devices. This makes it possible to identify health concerns or consequences early on and take appropriate action. Better chronic condition management, fewer hospital admissions, and lower healthcare expenses are all possible with AI-powered RPM [9, 10]. AI algorithms

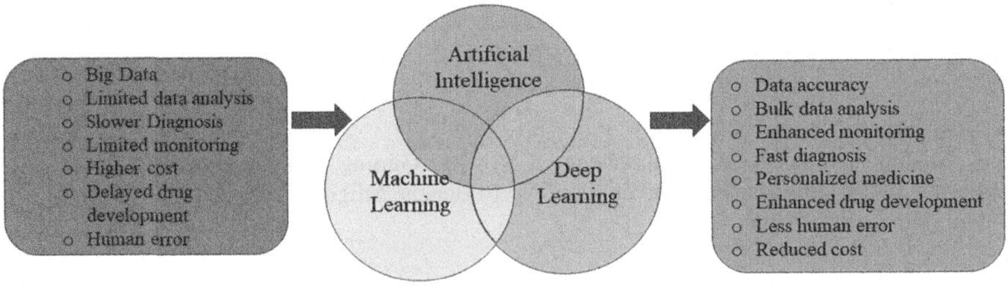

Fig. 1 An overview of AI in healthcare

can evaluate medical images generated from tests such as CT scans, X-rays, MRI, etc., identify subtle trends, and offer impartial, reliable evaluations. These techniques improve patient monitoring, treatment planning, and disease identification in various medical specializations [11]. Additionally, it can automate procedures like robotic surgery and medical coding, enhancing surgeons' strength and range of motion [12]. AI technologies and algorithms can optimize supply chain management and administrative rules, improving patient flow and decreasing wait times. They can also estimate patient intake and identify peak operation hours. AI-powered scheduling programs that maximize equipment usage and staff rosters. AI also improves the administration of medical records by automating tedious administrative tasks [13].

While AI offers many benefits and has contributed to the development of the healthcare industry, it also faces significant obstacles, including incomplete, heterogeneous, and unclear data sets. The aim is to customize treatments, medical decisions, and health procedures for each patient; nevertheless, this calls for integrating, fusion, and mapping of various data. To ensure meaningful outcomes, explainable AI must consider multiple data sources and enable medical practitioners to comprehend how machines make decisions [1]. Concerns about data quality, privacy, and the potential for AI to instill bias in people are additional challenges that need to be addressed. Thorough validation is necessary before integrating AI technology into clinical procedures. Despite these challenges, AI can transform medical diagnosis, improving patient outcomes and the efficiency of healthcare delivery. AI will likely play an increasing role in aiding and enabling physicians as technology advances. Considering this, this chapter aims to examine the benefits of artificial intelligence in healthcare while outlining areas for improvement.

2 AI in Disease and Diagnostics

Disease diagnosis accurately identifies a diseased state by closely examining the symptoms, test results, and history of previous diseased conditions. Various techniques like X-rays, CT scans, MRI, blood tests, biopsies, etc., are used to analyze the condition of the patients. It also aids in identifying the progress of medical treatment and detects any health condition before it becomes adverse. The healthcare professionals interpret the test results and help them identify the disease. However, misinterpretation of test results, shortage of experienced healthcare professionals, the inability of the diagnostic techniques for accurate prognosis of disease, etc., are some of the limitations of these techniques. Artificial Intelligence (AI) in disease diagnostics is the application of advanced computational methods and algorithms for accurate identification and prognosis of diseases. It encompasses various tools to analyze test results, identify biomarkers, and help healthcare professionals diagnose disease, from early detection to treatment. The introduction of AI improved the diagnostic field with significant prediction accuracy, speed, and efficiency. AI can analyze the medical images produced by X-rays, MRI, etc., and help healthcare professionals accurately detect disease. With the advanced algorithms developed, AI can quickly utilize large amounts of data to detect a person's disease state with high accuracy and efficiency [11, 14].

3 Types of AI Technologies Used in Diagnostics

ML, DL, Natural Language Processing (NLP), and computer vision are the different types of AI used in disease diagnosis. ML lets the machine learn from a large amount of data using mathematical and statistical approaches. No programming is required in this process [15]. ML can be classified into supervised and unsupervised learning. In supervised learning, the algorithm is learned from the labelled data. The model gets trained from the labelled dataset, and the training is used to predict the outcomes of new datasets. This technique will aid in predicting the diagnosis of a disease in a patient by looking at the patient's data. Algorithms like decision trees and support vector machines are used for these cases. It helps identify already-known diseases. In unsupervised learning, the algorithm is trained on unlabeled data. It is helpful when the experts cannot make any pattern from the data set. Hence, an unsupervised learning algorithm will aid in finding hidden patterns or intrinsic structures from the data set. It helps predict rare or uncommon diseases by looking at the unusual patterns in the patient dataset.

Neural Networks (NN) and DL are subfields of machine learning that utilize multiple layers to learn the data attributes. The number of layers applied to the neural network determines the depth of the deep learning process. Different neural network architectures have advanced, which include convolutional neural networks (CNNs), recurrent neural networks (RNNs), Transformers, and autoencoders (AEs) that have various applications in biology, including face identification, organ location, recognition of biological images, etc. [15, 16]. NLP is the field of AI that equips computers to understand, analyze, and generate human language using algorithms. NLP in disease diagnosis has wide applications as it can leverage the vast amount of unstructured data generated in healthcare [17]. Text data created in the medical field includes patient case sheets, reports of health records, discharge summaries, etc. NLPs use semantic analysis to convert human language into text data, understanding and interpreting the meaning of these texts and creating relationships between them [18]. Computer vision is another technique used by AI in the medical field. It is utilizing various algorithms to interpret and make decisions based on visual data [19]. This technique is instrumental in surgery and disease treatment. It uses pattern recognition techniques, including CNNs, image segmentation, and feature extraction, for adequate pathology detection, classification, and quantification in X-ray, CT, MRI, and histopathologic image [20–22].

4 Current Applications of AI in Disease Diagnosis

AI revolutionized disease diagnosis by enhancing accuracy, efficiency, and early detection across various medical fields. Its integration will help improve diagnostic capabilities, support personalized medicine, and better patient outcomes. AI is applied to multiple disease diagnoses, including cancers, cardiovascular diseases, and neurological diseases (Fig. 2). Cancer has become one of the most common causes of mortality. One of the best ways to accomplish reduced mortality is the early prediction of cancer onset. AI plays a multifaceted role in cancer therapy, from cancer detection to drug discovery and validation [23, 24]. AI algorithms analyze CT scans and chest X-rays to identify lung nodules that can indicate lung cancer. These algorithms can also help closely examine the nodules' size, shape, and growth for early detection and prediction of the state of malignancy in lung cancer [25]. AI algorithms classify mammograms to detect abnormalities, such as microcalcifications and masses, that may indicate the presence of breast cancer [26]. An ensemble deep learning algorithm can analyze the class and invasiveness stage of breast cancer from the whole slide image histopathology slide [24]. AI enhances 3D breast tomosynthesis by reconstructing three-dimensional pictures of the breast, thus allowing detailed study of the breast tissue and high-accuracy tumor detection [27, 28].

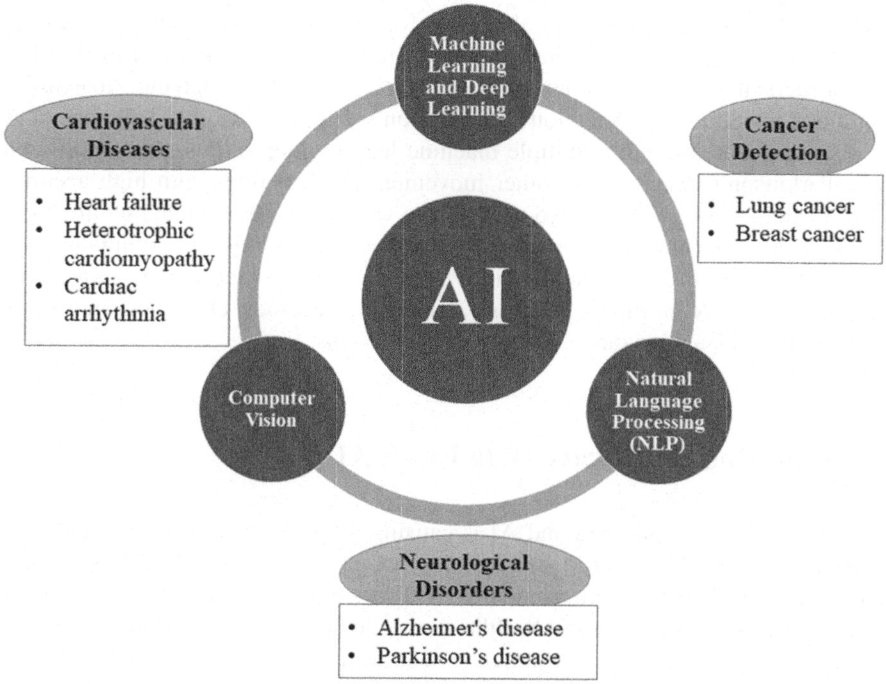

Fig. 2 Application of AI in disease diagnosis

The application of AI helps detect various cardiovascular diseases. AI algorithms run over patient medical records, symptoms, and imaging results to detect incipient heart failure, such as echocardiograms and other data sources. AI facilitates early intervention and treatment by subtle changes in cardiac function and structure [29]. AI can even detect Heterotrophic cardiomyopathy (HCM) from the deregulated electrocardiogram values. Machine learning and neural network-based learning were developed based on ECG data and are now widely applied in detecting cardiac arrhythmia in patients. The CNN algorithm of AI is validated for diagnosing and predicting the risk factors for patients with coronary artery disease [30]. AI-powered wearable devices and remote monitoring systems continuously collect and analyze data on heart rate variability, fluid status, and activity levels. AI algorithms detect deviations from standard patterns and detect potential worsening of heart failure symptoms much faster. This early detection can prevent hospitalizations and improve patient management [31, 32].

The usage of AI in neurological diseases helps early detection and prediction of biomarkers, radio imaging, etc. This spans areas from predicting a neurological condition or its severity to localized regions of interest, as has already been demonstrated in stroke diagnostics for lesion detection in diffusion-weighted MR imaging [33]. AI systems can analyze brain scans (MRI, PET) to discover early indications of Alzheimer's that humans may overlook. This includes identifying biomarkers such as amyloid plaques and tau tangles [34]. Cheung and colleagues created a retinal photograph-based deep learning algorithm for identifying Alzheimer's disease by looking at the various ocular imaging modalities. The algorithm can even differentiate persons with Alzheimer's and dementia and patients having Alzheimer's without dementia [35]. Detection of Parkinson's disease is another hallmark of AI in disease diagnosis. With multiple machine learning algorithms, AI can differentiate Parkinson's disease from other movement abnormalities with high accuracy [36]. Besides these, AI is applied to diagnose various diseases, including infectious diseases, develop personalized medicines, and perform dermatology, ophthalmology, gastroenterology, etc. These applications as shown in Fig. 2 highlight the transformative potential of AI in illness diagnosis, increasing accuracy, efficiency, and patient outcomes across a wide range of medical professions.

5 Analyzing Healthcare Data Using AI

The combination of big data and AI is causing a transformation in the healthcare industry (Fig. 3). AI-powered analytics are revolutionizing how healthcare data is transformed to better patient outcomes. It also strengthens modern healthcare by analyzing enormous amounts of healthcare data, including electronic health records, medical imaging, genomic data, and other healthcare information, to predict changes and reveal previously unknown patterns and insights. AI approaches can extract valuable information from big data in healthcare and utilize it to advance healthcare and transform many elements of patient care and administrative operations [37, 38]. NLP

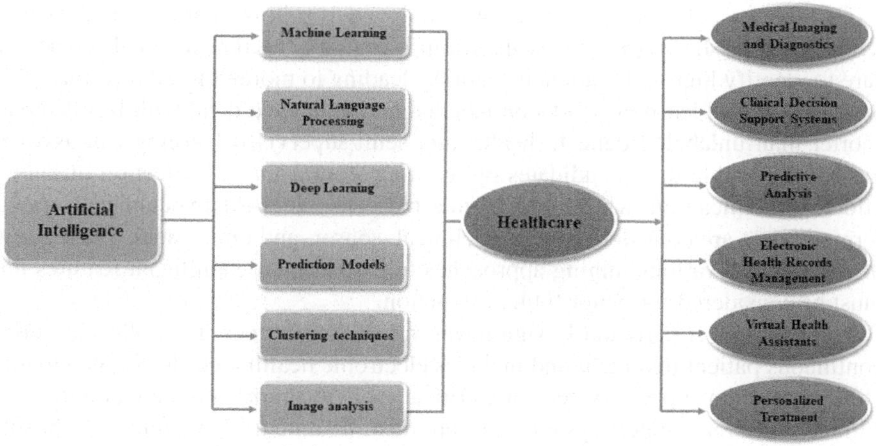

Fig. 3 Artificial Intelligence in healthcare data analysis

is becoming more popular in healthcare as health systems recognize its capability to search, analyze, and interpret large amounts of data. NLP is a subset of AI that allows machines to understand and communicate in natural language using various techniques such as ML and DL. The healthcare industry inputs a large volume of data into electronic health records daily, making it difficult for physicians to aggregate that critical data. However, NLP can help healthcare professionals identify hidden patterns in medical reports quickly and accurately, providing new perspectives on diseases or health issues that were previously unknown. The NLP algorithm can extract clinically significant information from unstructured medical data such as medical records, articles, and text-based data, allowing for finding relevant information faster and accelerating clinical trial matching [39, 40]. Moreover, the subdiscipline of AI, known as Machine Learning, is a collection of techniques used to address data analysis problems by identifying patterns in the data.

Machine learning can be used for diverse applications and is split into three divisions: supervised, unsupervised, and semi-supervised [41]. These three learning involve analyzing healthcare data in various ways, where supervised learning involves training a model on the data and using this trained model to make predictions on the new data, which may predict an individual's likelihood of developing diseases like cancer and heart diseases or diabetes, etc. based on factors like medical history, genetics, and lifestyle [42, 43]. Unsupervised learning is also used to train data when the target value is unknown. Unsupervised learning is primarily concerned with discovering patterns in data rather than making predictions. It is mostly utilized in anomaly detection, which can reveal medical errors, fraud, or early indicators of disease through patterns. However, generative models such as variational autoencoders can learn the underlying data distribution and make predictions without labelled data, which is important for tasks like disease progression prediction [44, 45].

Furthermore, clustering methods are employed, resulting in automatic data clustering based on shared traits. This algorithm is generally used in personalized healthcare to identify high-risk patient categories, leading to more tailored therapies [41]. Semi-supervised learning works on labelled and unlabeled data, with labelled data shorter than unlabeled data. In healthcare, semi-supervised learning can assist in uncovering viable drug candidates by learning from a limited selection of known effective chemicals and a huge pool of untested molecules [46]. In healthcare, several forms of data are considered, such as clinical, sensor, and omics data. These forms of data require various mining approaches to extract more significant features that must be considered for better future prediction.

Clinical data is particularly significant in healthcare since it is collected during continuous patient treatment and includes electronic health records. Related to this, clinical decision support systems are critical tools for improving patient outcomes. Clinical decision support systems can optimize treatment plans, minimize healthcare costs, and increase patient safety, which is strengthened by integrating AI [47, 48]. Various deep-learning models have assisted in identifying complicated pattern extraction from varied medical data, including pictures and sequential data such as ECGs, thereby improving diagnostic accuracy. Diagnostic support is one of the most essential applications for AI in healthcare. AI algorithms' intense learning models have demonstrated the ability to analyze medical imagery such as X-rays and MRIs and make correct diagnoses. AI in healthcare is revolutionizing several elements, including data entry and appointment scheduling, allowing healthcare providers to focus on patient care and provide higher-quality care while remaining cost-effective.

In addition to radiography, machine learning algorithms in medical image analysis have been greatly developed to diagnose or classify the severity of the condition [41, 48]. Several attempts have been made to construct AI systems based on electronic health records. Additionally, artificial intelligence promises a disruptive approach to population health management, with enormous potential to enhance healthcare outcomes for individuals and communities. Extending the program to analyze data in healthcare decreases resource usage, treatment costs, and time [49]. In addition, it reduces the time it takes to diagnose and make decisions, significantly impacting treatment plans and outcomes. Nowadays, people employ a variety of Internet of Things (IoT) devices to help them improve their health by gathering health information from everyday activities. AI can analyze sensor data from medical equipment to determine when maintenance is needed. It also allows for proactive maintenance and gives reminders for prescription visits [50].

Furthermore, AI models are utilized for drug discovery reasons, employing data from clinical trials and patient information, which are analyzed using machine learning approaches to find successful medications and pathways. Furthermore, a personalized medicine approach is developing using AI in healthcare, where AI can spot minor changes, leading to earlier detection and more accurate diagnosis [51, 52]. Although AI has proven more effective in drug repurposing due to its ability to evaluate massive volumes of data and discover unexpected patterns. This can be achieved using three major approaches: network-based, feature-based, and matrix-based, all focusing on collecting and using key information from various datasets to predict medication and disease connections. However, several difficulties need to be solved, such as data quality and security [53, 54]. Simply put, AI has been used for various reasons in data analysis as shown in Fig. 3, and it has the potential to assist in raising significant health issues.

6 Case Study

Sun et al. describe AI in cardiovascular diseases, specifying diagnostic and therapeutic aspects. AI algorithms have the potential to significantly speed up the detection and treatment of several cardiovascular disorders (CVD), including heart failure, atrial fibrillation, valvular heart disease, hypertrophic cardiomyopathy, and congenital heart disease. These algorithms focus on promoting stratification, typing, and outcome prediction of diseases by analysing huge amounts of healthcare data. Through techniques like deep learning and neural networks, AI is applied to devices for the early detection of CVDs, such as electrocardiograms (ECG) and echocardiograms. For instance, AI-ECG models have already been developed to screen conditions like valvular heart disease, atrial fibrillation, or hypertrophic cardiomyopathy with high accuracy and potential for early diagnosis. Moreover, it makes the role of the other auxiliary tools more effective, including computed tomography angiography (CCTA), and provides more accurate risk stratification of CVD patients, thereby improving personalized medicine. Models implemented with AI in this tenet are utilized for outcome prediction of CVD patients, including mortality prediction and prognostic assessment in coronary artery disease (CAD) patients. AI, therefore, holds great promise to be one day's game changer in cardiovascular medicine for the way diagnosis and therapy are conducted to improve patient care and outcomes [55].

The review by Marietta and colleagues on the systematic search for studies using AI in pathology microbiology specimens includes 110 articles after reviewing an initial pool of 4596. The review describes that, notwithstanding the victories won by most infections, infectious diseases remain a global health concern, specifically in resource-limited countries; accurate diagnosis remains essential for effective patient and public health management. Traditional microbe identification methods are time-consuming; for example, the manual microscopic examination requires skilled personnel. AI, particularly DL algorithms and CNN is very promising for the automation and accuracy of microorganism detection methods. AI applications on digital images taken through microscope cameras or smartphones have obtained highly satisfactory results. This, therefore, means that AI can be used as a reliable tool in aiding the pathologist in detecting microorganisms, showing potential benefits for developing countries where a skilled laboratorian and lack of equipment are prevalent. This review article depicts the potential of AI to enhance infectious disease diagnostics through automation, striving for objectivity in the interpretation and quantification of data. It is being demanded not just for an accessible database but also for technological development in such a new area, which will immensely benefit resource-constrained countries [56].

Nam et al. describe the status of AI, more specifically, ML/DL application in hepatology, focusing on its improvement in diagnostics, prognostics, and response prediction for liver diseases. Hepatology is a heterogeneous field, calling for integrating diverse data modalities to base diagnosis and treatment. AI has enabled advances by extracting therapeutically useful information from complicated data sets and AI systems that are decision-support aids in hepatology. However, there is still some retrogression to the fact that more clinical validation and regulatory approval are yet to be achieved. AI models can indirectly measure steatosis, inflammation, hepatocellular ballooning, and liver fibrosis staging in Non-alcoholic fatty liver disease and non-alcoholic steatohepatitis through a histopathological diagnosis. Development of AI models predicting clinical endpoints directly from histopathology images includes overall survival after resection in patients with (hepatocellular carcinoma) HCC, early recurrence after resection, and risk for HCC recurrence following liver transplantation. Abdominal CT delineation of the whole liver and liver lesions uses CNNs of U-Net architecture. This would be an essential step in surgical planning and treatment assessment. Histopathology, radiography, and clinical data are increasingly used in training models to predict outcomes and the likelihood of HCC recurrence following liver transplantation [57].

Mun et al. explored the transformative potential of AI in radiology and identified three critical pathways beyond current CNN capabilities. It has been said that AI efforts should refocus on the differences in radiology workflow and medical error reduction if they are to lead to demands and increased adoption within the radiology community. Proposed pathways include enhancements in CAD tools, optimizing radiology service productivity using AI-assisted workflow, and developing radionics for integrated diagnostics. With advancements in CAD tools, improvement in workflow efficiency, and data integration from radiology, pathology, and genomics, AI can potentially revolutionize radiology services and help facilitate personalized precision medicine [58].

According to Zhang et al., oncology has implemented AI for tumor diagnosis, prognosis prediction, and treatment administration. In combination, developing DL algorithms Marvin and ML has significantly increased the accuracy of cancer detection and personalized treatment strategies. The study finds that AI has played a significant role in furthering cancer research and clinical practice. AI technologies, such as ChatGPT and DL models, can be harnessed to aid health professionals in decision-making for more exact diagnoses and appropriate treatment plans against cancer. DL models like CNN are applied for image-based tumor screening, which helps precisely diagnose, stage, and grade tumors. Machine learning algorithms such as random forest, naive Bayes, k-nearest neighbor, and support vector machines detect cancer early on. These algorithms improve the accuracy of screening tests to arrive at an early diagnosis and treatment. Further, deep learning models from the reference datasets allow robust somatic mutation calling in cancer samples. This model thus differentiates between somatic and germline origin of genomic alterations, hence beneficial for personalized therapy in cancer. The abundance of state-of-the-art Artificial Intelligence approaches has been utilized in oncology for cancer detection, prognosis, and treatment administration. CNNs dominate with great promise in the medical imaging data analysis arenas in mammograms, magnetic resonance imaging scans, computed tomography scans, etc., for early detection and diagnosis of cancerous lesions; opinion CDSS like Watson for Oncology uses AI algorithms to help treatment decisions by health professionals based on analysis of patient data and recommendation of the most suitable options based on evidence-based guidelines [59]. Apart from the involvement of AI in the healthcare system, the computational approaches also came as new hope in terms of vaccine development, drug design and novel inhibitor identification [60–65], which will also help improve healthcare systems' overall accuracy and enhancement.

7 Security, Privacy, and Ethical Implications of AI in Healthcare

Artificial intelligence will benefit patient care, diagnosis, treatment, and medical data analysis. AI has the potential to enhance and optimize medical procedures, leading to better patient outcomes and less data overload for medical staff. Healthcare data security, privacy, responsibility, and justice are ethical considerations that are becoming more pressing despite the expanding usage of artificial intelligence in the field. These concerns become more crucial when AI technology is applied in therapeutic settings [66, 67]. It is essential to comprehend these ramifications to deliver ethical and fair healthcare. AI produces massive volumes of patient data in healthcare, which needs to be secured against unwanted access. Cybersecurity attacks could jeopardize patient safety and data integrity in healthcare AI systems. The intricacy of AI algorithms might present dangers for the misuse of private medical data [68, 69]. Data security is also a major ethical concern because healthcare data is one of the most frequently targeted areas for identity fraud. Healthcare providers and AI developers must balance the need for innovation and the security of systems to protect all patient data and ensure data integrity [68, 70]. Employing biased data in applications regarding data privacy, justice, and fairness is also seen as unethical. Ethical considerations regarding liability in AI technology include holding healthcare providers, institutions, and manufacturers accountable for the procedure [70]. Moreover, data ownership is problematic when combining data from several sources for algorithmic training and decision-making. The ownership of data raises an ethical question. The lack of transparency of the multiple built-in algorithms raises many ethical issues. Finding the root cause of errors is challenging since patients and healthcare providers find it difficult to trust AI-driven advice [71]. It is critical to assess the ethical ramifications and protect patients' rights and interests as AI continues to transform healthcare systems across the globe.

8 Limitations of AI in Healthcare

Several restrictions are associated with developing the AI-based method for increasing procedure and accuracy; these restrictions primarily stem from data privacy, initial data collecting, and other factors. The dataset from the first sample that was obtained was needed by the AI-based application, and this is the essential foundation for the research. The obstacles to achieving accuracy are the complexity and redundancy in the sample collection and their initial data creation. However, the healthcare data include differences within the acquired information due to the initial patient lacking information and occasionally erroneous data collection. This is because the AI-based application primarily relies on the bulk of data to train the model and its validation. However, because AI-based solutions lack interpretability, the healthcare industry largely avoids proposals regarding confidence or reliability. Additionally, the advanced infrastructure needed for the application to improve healthcare may not be available, and compatibility issues may arise. Aside from that, the number of AI-based methods was growing daily. The main concerns, however, are the security and privacy of the acquired sample and the corresponding personal information, ethical considerations, a constrained environment (depending on the collected sample), issues with trust, and technological limits.

9 Advancement and Challenges

With the increasing number of new diseases, the diagnosis and treatment process should also level up accordingly to overcome the increase in mortality rate. The introduction of AI is a transforming factor in healthcare. AI has emerged as a disruptive force in the healthcare sector, with the potential to advance therapy, drive innovation, and make significant promises for the present and future diagnosis of diseases. AI utilizes algorithms like ML and DL, Where DL imitates human brain networks to find intricate links in data, and machine learning uses data to find patterns. These technologies, both subsets of artificial intelligence, can identify relationships far more quickly and precisely to surpass human intelligence and diagnose, analyze and treat a condition. Overall, AI's main aim in healthcare is to analyze large data efficiently and at a much faster rate for early diagnosis of conditions, to provide correct treatment, and to predict the patient's outcome post-treatment.

AI has multiple advantages in the medical and healthcare fields; one of the most important is disease detection. AI can analyze data from enormous datasets from all over the available information, which helps analyze the symptoms and progression to interpret or diagnose accurately and quickly. In the field of diagnostics, AI has demonstrated incredible proficiency at interpreting X-rays, CT scans, and MRI pictures to diagnose diseases and abnormalities; similarly, AI medical diagnostic technologies can also be beneficial for oncological scans to detect tumors and differentiate between different types additionally, research indicates that deep learning systems are much better than radiologists in terms of accuracy. Early accurate detection leads to early treatment of the condition, increasing survival chances. AI uses algorithms like ML, which can analyze massive patient data sets, including medical histories, imaging scans, and personal histories, to quickly spot patterns and abnormalities that would be difficult for human eyes to see otherwise. This makes it possible for medical professionals to identify patients more precisely and provide individualized treatment regimens.

With the advancement of tailored diagnosis, recommended personalized medicine or treatment can have a great impact on the patient's outcome; personalized medicine treatments consider factors like patients' lifestyle, gender, age, occupation, and most importantly, the genetic makeup based on all the large data of the patient, most accurate diagnosis and treatment can be provided. AI can also detect risk factors linked to diseases and analyze patient outcomes after the treatments. ML algorithms are used to predict the response to treatments and medicines, as well as the survival rate and the course of illness; with these, medical professionals can gain from early intervention and proactive patient management. When making decisions, AI analyses the patient's data's possible outcomes and is not manipulated by emotions or human bias. It solely depends on the data and algorithms, and an unbiased decision is made, ultimately resulting in better patient health outcomes.

AI has also impacted drug discovery and advancement with the use of algorithms ML and DL; it can quickly and efficiently analyze the data, which in turn accelerates the process of identifying the possible targets that are effective and helps the researchers to save time and expense and the possibility of error is most likely to be less. AI also makes drug repurposing possible with its capability of analyzing enormous amounts of data in comparatively little time, presenting a viable method for identifying novel therapeutic uses for currently available medications. This not only drastically cuts down on the time and expenses involved in traditional drug development but also makes it possible to identify novel applications for already-established pharmaceuticals. The introduction of AI in hospitals can use algorithms to examine medical data and spot inefficiencies; early efficient detection of conditions can cut down the cost of unnecessary examinations and time and predict the patient's possible outcomes. Therefore, they can cut waste and simplify procedures to minimize administrative burden and improve cost-effectiveness.

While AI has much potential to improve healthcare, many issues with data, openness, regulation, understanding, and implementation must be resolved before this game-changing technology's full benefits. AI requires a huge amount of personalized patient data to analyze, which can be of great concern for the patient; it is critical to guarantee that personal data is protected from unauthorized access and improper use. AI is also called the 'black box' because it might be challenging to decipher how an AI system arrived at a given choice; healthcare experts find it difficult to trust the AI due to its lack of openness fully. Every data set has some degree of bias based on gender, sexual orientation, race, sociologic, environmental, or economic characteristics; the most frequent ethical problem is unfairness resulting from bias in data sources. AI is a newly developed technology, and not all healthcare workers are aware of it, so implementing AI in healthcare sectors will require some changes that might be challenging to adapt. Also, successfully maintaining an AI pilot is difficult as it requires much-sustained effort and financial support.

10 Conclusion

At present, along with the emerging and re-emerging pathogens, the complications are also getting complex, and to handle this situation, AI emerged as a new hope to classify the data and information, analyze information, predict the possibilities, and advance the healthcare system. The involvement of AI, followed by their subset of algorithms, is advancing daily and setting a significant path that can easily handle uncontrolled situations. Moreover, to extract accurate information, advancements and early precise data collection are essential to get better insight. Furthermore, updating the algorithm and creating a new pipeline to get the curated data information is required, followed by easy setup and easy handling, which can be helpful in the healthcare system.

References

1. Holzinger, A., Langs, G., Denk, H., Zatloukal, K., Müller, H.: Causability and explainability of artificial intelligence in medicine. Wiley Interdisc. Rev.: Data Mining Knowl. Discovery **9**, e1312 (2019)
2. Tariq, M., Hayat, Y., Hussain, A., Tariq, A., Rasool, S. Principles and perspectives in medical diagnostic systems employing artificial intelligence (AI) algorithms. Int. Res. J. Econ. Manage. Stud. IRJEMS **3** (2024)
3. Ghaffar Nia, N., Kaplanoglu, E., Nasab, A.: Evaluation of artificial intelligence techniques in disease diagnosis and prediction. Discover Artif. Intell. **3**, 5 (2023).
4. Tekkeşin, A.İ: Artificial intelligence in healthcare: past, present and future. Anatol. J. Cardiol. **22**, 8–9 (2019)
5. Tariq, M., Hayat, Y., Hussain, A., Tariq, A., Rasool, S.: Principles and perspectives in medical diagnostic systems employing artificial intelligence (AI) algorithms. Int. Res. J. Econ. Manage. Stud. IRJEMS 3
6. Baker, A., Perov, Y., Middleton, K., Baxter, J., Mullarkey, D., Sangar, D., Butt, M., DoRosario, A., Johri, S.: A comparison of artificial intelligence and human doctors for the purpose of triage and diagnosis. Front. Artif. Intell. **3**, 543405 (2020)
7. Khalifa, M., Albadawy, M.: AI in diagnostic imaging: Revolutionising accuracy and efficiency, Comput. Methods Programs Biomed. Update 100146 (2024)
8. Pinto-Coelho, L.: How artificial intelligence is shaping medical imaging technology: a survey of innovations and applications. Bioengineering (Basel) 10 (2023)
9. Devi, K.J., Alghamdi, W., Divya, N., Alkhayyat, A., Sayyora, A., Sathish, T.: Artificial Intelligence in healthcare: diagnosis, treatment, and prediction. In: E3S Web of Conferences, EDP Sciences, p. 04043 (2023)
10. Yelne, S., Chaudhary, M., Dod, K., Sayyad, A., Sharma, R.: Harnessing the power of AI: a comprehensive review of its impact and challenges in nursing science and healthcare. Cureus **15**, e49252 (2023)
11. Al-Antari, M.A.: Artificial intelligence for medical diagnostics-existing and future AI technology! Diagnostics (Basel) **13** (2023)

12. Reddy, K., Gharde, P., Tayade, H., Patil, M., Reddy, L.S., Surya, D.: Advancements in robotic surgery: a comprehensive overview of current utilizations and upcoming frontiers. Cureus **15**, e50415 (2023)
13. Helo, P., Hao, Y.: Artificial intelligence in operations management and supply chain management: an exploratory case study. Prod. Plan. Control **33**, 1573–1590 (2022)
14. Mishra, S.K., Jeba Praba, J., Georrge, J.J.: An emerging trends of bioinformatics and big data analytics in healthcare. In: Digital Transformation in Healthcare 5.0: Volume 2: Metaverse, Nanorobots and Machine Learning, vol. 159 (2024)
15. Ahsan, M.M., Luna, S.A., Siddique, Z.: Machine-learning-based disease diagnosis: a comprehensive review. Healthcare (Basel) **10** (2022)
16. Sapoval, N., Aghazadeh, A., Nute, M.G., Antunes, D.A., Balaji, A., Baraniuk, R., Barberan, C.J., Dannenfelser, R., Dun, C., Edrisi, M., Elworth, R.A.L., Kille, B., Kyrillidis, A., Nakhleh, L., Wolfe, C.R., Yan, Z., Yao, V., Treangen, T.J.: Current progress and open challenges for applying deep learning across the biosciences. Nat. Commun. **13**, 1728 (2022)
17. Hao, T., Huang, Z., Liang, L., Weng, H., Tang, B.: Health natural language processing: methodology development and applications. JMIR Med. Inform. **9**, e23898 (2021)
18. Velupillai, S., Mowery, D., South, B.R., Kvist, M., Dalianis, H.: Recent advances in clinical natural language processing in support of semantic analysis. Yearb. Med. Inform. **10**, 183–193 (2015)
19. Esteva, A., Chou, K., Yeung, S., Naik, N., Madani, A., Mottaghi, A., Liu, Y., Topol, E., Dean, J., Socher, R.: Deep learning-enabled medical computer vision. NPJ Digit. Med. **4**, 5 (2021)
20. Gao, J., Yang, Y., Lin, P., Park, D.S.: Computer vision in healthcare applications. J. Healthc. Eng. **2018**, 5157020 (2018)
21. Hashimoto, D.A., Rosman, G., Rus, D., Meireles, O.R.: Artificial intelligence in surgery: promises and perils. Ann. Surg. **268**, 70–76 (2018)
22. Johnson, K.B., Wei, W.Q., Weeraratne, D., Frisse, M.E., Misulis, K., Rhee, K., Zhao, J., Snowdon, J.L.: Precision medicine, AI, and the future of personalized health care. Clin. Transl. Sci. **14**, 86–93 (2021)
23. Guren, M.G.: The global challenge of colorectal cancer. Lancet Gastroenterol. Hepatol. **4**, 894–895 (2019)
24. Balasubramanian, A.A., Al-Heejawi, S.M.A., Singh, A., Breggia, A., Ahmad, B., Christman, R., Ryan, S.T., Amal, S.: Ensemble deep learning-based image classification for breast cancer subtype and invasiveness diagnosis from whole slide image histopathology. Cancers **16**, 2222 (2024)
25. Chiu, H.Y., Chao, H.S., Chen, Y.M.: Application of artificial intelligence in lung cancer. Cancers (Basel) **14** (2022)
26. Lamb, L.R., Lehman, C.D., Gastounioti, A., Conant, E.F., Bahl, M.: Artificial intelligence (AI) for screening mammography, from the AJR special series on AI applications. AJR Am. J. Roentgenol. **219**, 369–380 (2022)
27. Park, E.K., Kwak, S., Lee, W., Choi, J.S., Kooi, T., Kim, E.-K.: Impact of AI for digital breast tomosynthesis on breast cancer detection and interpretation time. Radiol.: Artif. Intell. **6** e230318 (2024)
28. Letter, H., Peratikos, M., Toledano, A., Hoffmeister, J., Nishikawa, R., Conant, E., Shisler, J., Maimone, S., Diaz de Villegas, H.: Use of artificial intelligence for digital breast tomosynthesis screening: a preliminary real-world experience. J. Breast Imaging **5**, 258–266 (2023)
29. Somani, S.S., Narayan, S.M., Rogers, A.J.: Artificial intelligence in cardiac electrophysiology. In: Artificial Intelligence in Clinical Practice, pp. 475–496. Elsevier (2024)
30. Haq, I.U., Chhatwal, K., Sanaka, K., Xu, B.: Artificial Intelligence in cardiovascular medicine: current insights and future prospects. Vasc. Health Risk Manag. **18**, 517–528 (2022)
31. Muhammad, Y., Tahir, M., Hayat, M., Chong, K.T.: Early and accurate detection and diagnosis of heart disease using intelligent computational model. Sci. Rep. **10**, 19747 (2020)
32. Baxt, W.G., Shofer, F.S., Sites, F.D., Hollander, J.E.: A neural network aid for the early diagnosis of cardiac ischemia in patients presenting to the emergency department with chest pain. Ann. Emerg. Med. **40**, 575–583 (2002)

33. Segato, A., Marzullo, A., Calimeri, F., De Momi, E.: Artificial intelligence for brain diseases: a systematic review. APL Bioeng **4**, 041503 (2020)
34. Zhang, W., Li, Y., Ren, W., Liu, B.: Artificial intelligence technology in Alzheimer's disease research. Intractable Rare Dis. Res. **12**, 208–212 (2023)
35. Cheung, C.Y., Ran, A.R., Wang, S., Chan, V.T.T., Sham, K., Hilal, S., Venketasubramanian, N., Cheng, C.Y., Sabanayagam, C., Tham, Y.C., Schmetterer, L., McKay, G.J., Williams, M.A., Wong, A., Au, L.W.C., Lu, Z., Yam, J.C., Tham, C.C., Chen, J.J., Dumitrascu, O.M., Heng, P.A., Kwok, T.C.Y., Mok, V.C.T., Milea, D., Chen, C.L., Wong, T.Y.: A deep learning model for detection of Alzheimer's disease based on retinal photographs: a retrospective, multicentre case-control study. Lancet Digit. Health **4**, e806–e815 (2022)
36. Dennis, A.P., Strafella, A.P.: The role of AI and machine learning in the diagnosis of Parkinson's disease and atypical parkinsonisms. Parkinsonism Relat Disord 106986 (2024)
37. Secinaro, S., Calandra, D., Secinaro, A., Muthurangu, V., Biancone, P.: The role of artificial intelligence in healthcare: a structured literature review. BMC Med. Inform. Decis. Mak. **21**, 125 (2021)
38. Albahri, A.S., Duhaim, A.M., Fadhel, M.A., Alnoor, A., Baqer, N.S., Alzubaidi, L., Albahri, O.S., Alamoodi, A.H., Bai, J., Salhi, A.: A systematic review of trustworthy and explainable artificial intelligence in healthcare: assessment of quality, bias risk, and data fusion. Inf. Fusion **96**, 156–191 (2023)
39. Le Glaz, A., Haralambous, Y., Kim-Dufor, D.H., Lenca, P., Billot, R., Ryan, T.C., Marsh, J., DeVylder, J., Walter, M., Berrouiguet, S., Lemey, C.: Machine learning and natural language processing in mental health: systematic review. J. Med. Internet Res. **23**, e15708 (2021)
40. Patra, B.G., Sharma, M.M., Vekaria, V., Adekkanattu, P., Patterson, O.V., Glicksberg, B., Lepow, L.A., Ryu, E., Biernacka, J.M., Furmanchuk, A., George, T.J., Hogan, W., Wu, Y., Yang, X., Bian, J., Weissman, M., Wickramaratne, P., Mann, J.J., Olfson, M., Campion, T.R., Weiner, M., Pathak, J.: Extracting social determinants of health from electronic health records using natural language processing: a systematic review. J. Am. Med. Inform. Assoc. **28**, 2716–2727 (2021)
41. An, Q., Rahman, S., Zhou, J., Kang, J.J.: A comprehensive review on machine learning in healthcare industry: classification, restrictions, opportunities and challenges. Sensors (Basel) 23 (2023)
42. Gu, X., Deligianni, F., Han, J., Liu, X., Chen, W., Yang, G.Z., Lo, B.: Beyond supervised learning for pervasive healthcare. IEEE Rev. Biomed. Eng. **17**, 42–62 (2024)
43. Devi, K.R., Suganyadevi, S., Balasamy, K.: Healthcare data analysis using deep learning paradigm. Deep learning for cognitive computing systems: technological advancements and applications. De Gruyter 129–148 (2023)
44. Eckhardt, C.M., Madjarova, S.J., Williams, R.J., Ollivier, M., Karlsson, J., Pareek, A., Nwachukwu, B.U.: Unsupervised machine learning methods and emerging applications in healthcare. Knee Surg. Sports Traumatol. Arthrosc. **31**, 376–381 (2023)
45. Riazi, K., Ly, M., Barty, R., Callum, J., Arnold, D.M., Heddle, N.M., Down, D.G., Sidhu, D., Li, N.: An unsupervised learning approach to identify immunoglobulin utilization patterns using electronic health records. Transfusion **63**, 2234–2247 (2023)
46. Sahu, H., Kashyap, R., Dewangan, B.K.: Hybrid deep learning based semi-supervised model for medical imaging. In: 2022 OPJU International Technology Conference on Emerging Technologies for Sustainable Development (OTCON), pp. 1–6. IEEE (2023)
47. Karalis, V.D.: The integration of artificial intelligence into clinical practice. Appl. Biosci. **3**, 14–44 (2024)
48. Ramgopal, S., Sanchez-Pinto, L.N., Horvat, C.M., Carroll, M.S., Luo, Y., Florin, T.A.: Artificial intelligence-based clinical decision support in pediatrics. Pediatr. Res. **93**, 334–341 (2023)
49. Adamson, B., Waskom, M., Blarre, A., Kelly, J., Krismer, K., Nemeth, S., Gippetti, J., Ritten, J., Harrison, K., Ho, G., Linzmayer, R., Bansal, T., Wilkinson, S., Amster, G., Estola, E., Benedum, C.M., Fidyk, E., Estevez, M., Shapiro, W., Cohen, A.B.: Approach to machine learning for extraction of real-world data variables from electronic health records. Front. Pharmacol. **14**, 1180962 (2023)

50. Subhan, F., Mirza, A., Su'ud, M.B.M., Alam, M.M., Nisar, S., Habib, U., Iqbal, M.Z.: AI-enabled wearable medical internet of things in healthcare system: a survey. Appl. Sci. **13** 1394 (2023)
51. Blanco-Gonzalez, A., Cabezon, A., Seco-Gonzalez, A., Conde-Torres, D., Antelo-Riveiro, P., Pineiro, A., Garcia-Fandino, R.: The role of AI in drug discovery: challenges, opportunities, and strategies. Pharmaceuticals (Basel) **16** (2023)
52. Mak, K.-K., Wong, Y.-H., Pichika, M.R.: Artificial intelligence in drug discovery and development. In: Drug Discovery and Evaluation: Safety and Pharmacokinetic Assays, pp. 1–38 (2023)
53. Al Kuwaiti, A., Nazer, K., Al-Reedy, A., Al-Shehri, S., Al-Muhanna, A., Subbarayalu, A.V., Al Muhanna, D., Al-Muhanna, F.A.: A review of the role of artificial intelligence in healthcare. J. Personal. Med. **13** 951 (2023)
54. Khanna, A., Jain, S.: Personalized drug treatment: transforming healthcare with AI. In: Artificial Intelligence and Machine Learning in Drug Design and Development, pp. 295–319
55. Sun, X., Yin, Y., Yang, Q., Huo, T.: Artificial intelligence in cardiovascular diseases: diagnostic and therapeutic perspectives. Eur. J. Med. Res. **28**, 242 (2023)
56. Marletta, S., L'Imperio, V., Eccher, A., Antonini, P., Santonicco, N., Girolami, I., Dei Tos, A.P., Sbaraglia, M., Pagni, F., Brunelli, M., Marino, A., Scarpa, A., Munari, E., Fusco, N., Pantanowitz, L.: Artificial intelligence-based tools applied to pathological diagnosis of microbiological diseases. Pathol. Res. Pract. **243**, 154362 (2023)
57. Nam, D., Chapiro, J., Paradis, V., Seraphin, T.P., Kather, J.N.: Artificial intelligence in liver diseases: improving diagnostics, prognostics and response prediction. JHEP Rep **4**, 100443 (2022)
58. Mun, S.K., Wong, K.H., Lo, S.B., Li, Y., Bayarsaikhan, S.: Artificial intelligence for the future radiology diagnostic service. Front. Mol. Biosci. **7**, 614258 (2020)
59. Zhang, C., Xu, J., Tang, R., Yang, J., Wang, W., Yu, X., Shi, S.: Novel research and future prospects of artificial intelligence in cancer diagnosis and treatment. J. Hematol. Oncol. **16**, 114 (2023)
60. Mishra, S.K., Priya, P., Rai, G.P., Haque, R., Shanker, A.: Coevolution based immunoinformatics approach considering variability of epitopes to combat different strains: a case study using spike protein of SARS-CoV-2. Comput. Biol. Med. **163**, 107233 (2023)
61. Vinjoda, P., Mishra, S.K., Sharma, K., Georrge, J.J.: In silico identification of novel drug target and its natural product inhibitors for herpes simplex virus, pp. 377–383. Elsevier, Nanotechnology and In Silico Tools (2024)
62. Vakhariya Sakina, S., Mishra, S.K., Sharma, K., Georrge, J.J.: Designing of a novel curcumin analogue to inhibit mitogen-activated protein kinase: a cheminformatics approach. J. Phytonanotechnol. Pharmaceutical Sci. **3**, 37–47 (2023)
63. Kachhadiya, D.K., Pooja, K., Mishra, S.K., George, J.J.: In Silico based identification of novel inhibitors for selected MDR protein from shigella species: a validation through molecular docking analysis. Educ. Admin.: Theory Practice **30**, 309–316 (2024)
64. Vaghasia, V.V., Sharma, K., Mishra, S.K., Georrge, J.J.: In silico identification of natural product inhibitor for multidrug resistance proteins from selected gram-positive bacteria, pp. 309–317. Elsevier, Nanotechnology and In Silico Tools (2024)
65. Mishra, S.K., Georrge, J.J.: In Silico investigation of Epigallocatechin-3-Gallate and its derivatives against the VEGF-A cancerous protein through molecular docking and MD simulation studies (2023)
66. Elendu, C., Amaechi, D.C., Elendu, T.C., Jingwa, K.A., Okoye, O.K., John Okah, M., Ladele, J.A., Farah, A.H., Alimi, H.A.: Ethical implications of AI and robotics in healthcare: a review. Medicine (Baltimore) **102**, e36671 (2023)
67. Santosh, K., Gaur, L., Santosh, K., Gaur, L.: Privacy, security, and ethical issues. In: Artificial Intelligence and Machine Learning in Public Healthcare: Opportunities and Societal Impact, pp. 65–74 (2021)
68. Hlávka, J.P.: Security, privacy, and information-sharing aspects of healthcare artificial intelligence. In: Artificial Intelligence in Healthcare, pp. 235–270. Elsevier (2020)

69. Roy, S.: Privacy prevention of health care data using AI. J. Data Acquisit. Process. **37**, 769 (2022)
70. Adeniyi, A.O., Arowoogun, J.O., Okolo, C.A., Chidi, R., Babawarun, O.: Ethical considerations in healthcare IT: a review of data privacy and patient consent issues. World J. Adv. Res. Rev. **21**, 1660–1668 (2024)
71. Chiruvella, V., Guddati, A.K.: Ethical issues in patient data ownership. Interact J. Med. Res. **10**, e22269 (2021)

Saurav Kumar Mishra conducts research in the areas of miRNAs, immunoinformatics, computational biology, molecular biology, and host–pathogen interaction analysis. His work also includes AI-based research in the healthcare sector, leveraging artificial intelligence for advancements in medical systems.

Tabsum Chhetri is currently engaged in drug target identification, transcriptomics data analysis, and the discovery of novel inhibitors in pathogen-related research. She has a strong interest in genomics and its impact on disease understanding, applying advanced computational bioinformatics methods to explore these relationships.

Anagha Balakrishnan previously served as a Junior Research Fellow at the Female Reproduction and Metabolic Syndrome Laboratory at the Rajiv Gandhi Centre for Biotechnology. She is now pursuing her Ph.D. in the Department of Bioinformatics at the University of North Bengal, with a research focus on designing novel drug candidates to combat cancer.

Sneha Roy is a dedicated chemoinformatics student specializing in drug discovery, personalized drug development, network pharmacology, and plant-derived natural products. She employs advanced computational and bioinformatics tools to innovate in these areas.

Kusum Gurung a postgraduate student, is focused on pharmacogenomics, the role of AI in healthcare, healthcare automation, and personalized medicine. She is passionate about transforming healthcare with advanced technologies to improve diagnostics, treatment success rates, and patient outcomes.

J. Jeba Praba is an Associate Professor in the Department of Computer Science and Applications at Christ College, Rajkot, Gujarat, India. She is a prominent researcher in cloud security in Gujarat, with over 18 years of experience in teaching, research, and academic administration. Previously, she served as the Chief Coordinator for a UGC-sponsored diploma program in Mobile Applications.

John J. Georrge is an Associate Professor in the Department of Bioinformatics at the University of North Bengal, Darjeeling, West Bengal, India. With extensive experience as a resource person, he has contributed to over 65 national and international symposia, workshops, and seminars, and has published more than 100 research papers and book chapters in internationally reputed journals and publications.

Exploring the Impact of Perceived Entrepreneurial Mindset and AI Adoption on Customer Experiences: A Study of Just Ask Chatbot Users

Shumayela Hasan and Richi Simon

Abstract This chapter examines the intersection of entrepreneurial mindset, AI adoption, and customer experience within the healthcare sector, focusing on the Just Ask Chatbot. Through a mixed-methods study involving 58 users, data were collected via purposive and snowball sampling, with structured interviews for qualitative insights. Quantitative analysis using Structural Equation Modeling (SEM) was conducted through Smart PLS SEM 4 software, while NVivo 13 was used for qualitative content analysis. Findings reveal a positive correlation between entrepreneurial mindset and customer experience, suggesting that proactive entrepreneurial attitudes play a critical role in enhancing customer interactions. Additionally, an entrepreneurial mindset positively influenced AI innovation adoption, emphasizing that entrepreneurs are inclined to integrate AI technologies within startup environments. Notably, ethical considerations emerged among users, reflecting the importance of addressing privacy and security challenges in AI adoption. The results underscore AI's capacity to elevate customer satisfaction, offering actionable insights for startup founders and industry leaders in promoting innovation-driven, ethically aware environments that leverage AI to achieve sustainable growth and competitive advantage.

Keywords Entrepreneurial mindset · AI adoption · Customer experience · Healthcare chatbots · Structural equation modeling · Content analysis · Innovation in startups · Digital transformation · Ethical concerns in AI · Sustainable growth

S. Hasan
The Bal Bhawan School, Bhopal, India
e-mail: aalifhasan7@gmail.com

R. Simon (✉)
The Bhopal School of Social Sciences, Bhopal, India
e-mail: richi_1988@yahoo.co.in

N. R. Vajjhala et al. (eds.), *Artificial Intelligence in Healthcare Information Systems—Security and Privacy Challenges*, Information Systems Engineering and Management 34, https://doi.org/10.1007/978-3-031-84404-1_6

1 Introduction

1.1 Artificial Intelligence (AI) and Healthcare

AI is revolutionizing the healthcare industry by enhancing diagnostic accuracy, streamlining administrative processes, and personalizing patient care. Machine learning algorithms analyze vast amounts of medical data, aiding in early disease detection and treatment planning. AI-driven tools, such as predictive analytics and natural language processing, improve patient outcomes by identifying potential health risks and optimizing clinical workflows. Additionally, AI supports telemedicine and remote monitoring, expanding access to healthcare services and enabling real-time patient monitoring. Overall, AI's integration in healthcare promises increased efficiency, reduced costs, and improved patient experiences.

AI in healthcare startups is revolutionizing the industry by enhancing diagnostics, treatment plans, and patient care management. These startups leverage machine learning algorithms and data analytics to predict disease outbreaks, personalize medicine, and streamline clinical workflows. They are also improving operational efficiencies in hospitals through predictive maintenance of medical equipment and optimizing resource allocation. Additionally, AI-powered virtual assistants and chatbots are enhancing patient engagement by providing 24/7 support and reducing the burden on healthcare professionals. By integrating AI, these startups are not only driving innovation but also making healthcare more accessible, efficient, and effective.

In practical terms, AI's role in diagnostics includes interpreting medical images, such as X-rays and MRIs, with a precision that rivals and sometimes surpasses human experts. For treatment plans, AI can recommend personalized therapies based on a patient's genetic profile and medical history, ensuring more effective and tailored care. AI also aids in drug discovery by identifying potential compounds and predicting their efficacy and safety, significantly accelerating the development process.

In administrative domains, AI automates routine tasks such as appointment scheduling, billing, and patient record management, freeing up healthcare workers to focus on patient care. Predictive analytics forecast patient admission rates and disease outbreaks, allowing healthcare providers to allocate resources more efficiently and prepare for potential surges in demand.

AI also enhances patient engagement and self-care through virtual health assistants and chatbots, which provide medical advice, track patient health metrics, and remind patients to take medications or follow treatment regimens. These tools are particularly beneficial for chronic disease management, where continuous monitoring and timely interventions are crucial.

The impact of AI is particularly evident in healthcare startups, which are at the forefront of integrating these advanced technologies into everyday medical practice. By harnessing AI, these startups are not only pioneering innovative solutions but also democratizing healthcare, making high-quality medical services more accessible to underserved populations. The continuous evolution of AI promises to transform healthcare into a more proactive, predictive, and personalized industry, ultimately leading to better health outcomes and a more efficient healthcare system. In today's fast-paced digital landscape, online startups are challenged not only to innovate but also to enhance customer experiences amidst intense competition. This study investigates the relationship between perceived entrepreneurial mindset, AI adoption, and its impact on customer experiences among 58 customers. Inspired by E.M. Rogers' Diffusion of Innovation Theory [1], we explore how entrepreneurial traits like risk-taking influence the integration of AI technologies. Using purposive sampling and a Mixed Method Approach, we uncover how proactive entrepreneurs leverage AI for growth. Preliminary findings highlight the positive correlation between perceived entrepreneurial mindset and AI adoption, emphasizing the role of AI in improving customer engagement and loyalty. This chapter advances academic understanding and offers practical insights for startup owners, policymakers, and industry stakeholders. It advocates for fostering environments conducive to entrepreneurial experimentation and strategic AI adoption to drive sustainable growth in the digital era.

1.2 Healthcare Scenario in India

The healthcare landscape in India presents a complex interplay of socio-economic and cultural factors that result in significant gender disparities. One of the most critical issues is the inequitable allocation of healthcare resources, where Indians spend less on female healthcare than on male healthcare. This discrepancy is not merely a result of differing healthcare needs between genders but stems from deeper systemic biases. According to Saikia and others, the gender gap in healthcare expenditure is not primarily driven by factors that typically influence healthcare spending but by underlying societal norms and practices that prioritize male health over female health [2].

Further exacerbating this inequality, women in India often have less access to inpatient care, especially in scenarios where healthcare expenditures are distress-driven, relying heavily on informal and financially unsustainable strategies [3]. These limitations manifest in higher mortality rates for women and significantly reduced access to essential medical services. However, the impact of these disparities is not uniform across the country; there are significant variations in healthcare access and outcomes across different states [4]. This uneven distribution underscores the need for targeted, state-specific interventions that account for local healthcare dynamics and cultural nuances.

The challenges in improving healthcare quality are compounded by a lack of reliable data and technical difficulties in measuring quality. The Indian healthcare system, both in the public and private sectors, is engaged in ongoing efforts to enhance data quality and develop better metrics for assessing and improving care. As India faces a rapidly changing disease profile, with a rising burden of chronic illnesses, it is increasingly urgent for state and central governments to collaborate with researchers and healthcare agencies to address these quality challenges [5].

Moreover, the issue of women's autonomy in health-related decision-making remains a significant concern. According to the National Family Health Survey (NFHS-5), only 10% of women in India have the autonomy to make independent decisions regarding their own health, and alarmingly, 11% of women believe that marital violence is justified if they refuse sex with their husbands. This highlights the deeply entrenched gender norms that continue to restrict women's rights and access to healthcare. The survey also reveals that nearly half of all pregnancies in India are unplanned, reflecting a broader global trend and underscoring the need for improved access to reproductive health services and education [6].

1.3 Just Ask Chatbot Initiative

"Just Ask Chatbot" emerges as a groundbreaking initiative, revolutionizing the landscape of adolescent and young adult healthcare and empowerment. Developed to bridge the information gap on sensitive topics like sexual and reproductive health and rights (SRHR), this AI-powered chatbot offers a safe haven for users to seek knowledge, guidance, and support.

At its core, "Just Ask" embodies accessibility and reliability. Users can freely inquire about a myriad of topics, ranging from puberty and menstruation to gender identity and family planning. This unrestricted access to information fosters a sense of empowerment among adolescents, enabling them to make informed decisions about their health and well-being.

The interactive nature of the chatbot further amplifies its impact. Through engaging conversations, users are not only provided with accurate information but also receive personalized guidance tailored to their individual needs. The inclusion of two relatable role models, Dhruv and Drishti, adds a human touch to the experience, making users feel understood and supported throughout their journey. The chatbot offers a predefined list of questions that can be selected as depicted in Fig. 1. The chatbot also seeks feedback from its users in terms of 1 to 3 star rating and inquires whether it should contact the user back or not so that the user may not be disturbed unnecessarily later.

Moreover, "Just Ask" goes beyond just providing information; it serves as a gateway to essential support services. By linking users to Adolescent Friendly Health Clinics (AFHCs), government facilities, and helplines, the chatbot ensures that individuals have access to comprehensive care and assistance when needed.

The success of the "Just Ask" initiative lies in its multifaceted approach to addressing the needs of adolescents and young adults. By combining cutting-edge AI technology with empathetic communication and seamless access to resources, the chatbot stands as a beacon of empowerment, guiding users towards healthier, more informed futures [7].

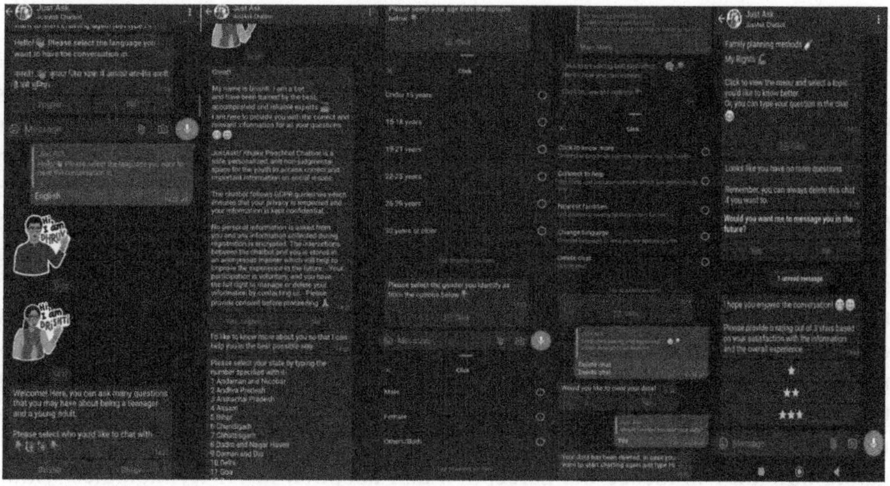

Fig. 1 Just Ask Chatbot interface

1.4 Just Ask Chatbot Ethical Compliances

Just Ask Chatbot can be accessed on whatsapp using the number—+918657024841 and simply typing "Hi". Thereafter the users get a choice of interacting with the two bots—Drishti and Dhruv. Just Ask Chatbot was developed with a strong emphasis on user data privacy and ethical considerations, in alignment with UNFPA's guide on the safe and ethical use of technology. This alignment suggests a deliberate effort to incorporate ethical principles into the chatbot's design. The broader ethical framework of UNFPA, which the chatbot follows, prioritizes safety and the protection of user rights, particularly in the context of using technology to combat gender-based violence and harmful practices. Key principles in this framework include ensuring safety, confidentiality, respect, non-discrimination, and a survivor-centered approach. Additionally, the chatbot is designed to provide a safe and non-judgmental space where users can access information about sexual and reproductive health and rights, reflecting a commitment to user well-being and autonomy. These ethical guidelines underscore the importance of privacy, safety, and creating a supportive environment for users. It is important to note that the bot seeks no personal information, but some demographic information is sought like geographical location, age and gender. After these basic queries are answered, the bot is all set to provide answers to a predefined set of questions. It also offers an option to delete data which can be used to clear the information added to the AI database (option shown in Fig. 1).

2 Theoretical Framework

This research aligns with the Diffusion of Innovation Theory by E.M. Rogers by framing its investigation within this theoretical framework. It explores how a perceived entrepreneurial mindset accelerates the adoption of AI innovations among startups, drawing parallels with the theory's emphasis on understanding adoption dynamics. The Diffusion of Innovation Theory identifies several key stages and factors that influence how and why innovations are adopted over time, including the characteristics of the adopters and the social systems involved. By focusing on the traits associated with innovators and early adopters, the study examines how customers perceive entrepreneurs with a proactive mindset in embracing AI technologies. Innovators and early adopters are often characterized by their willingness to take risks, openness to new ideas, and ability to understand and utilize new technologies ahead of the majority. The chapter examins these characteristics, investigating how they influence the perception and adoption of AI technologies within startup ecosystems. Additionally, the research investigates the consequences of AI adoption on customer experiences, extending the theory's focus on the outcomes of innovation diffusion. This includes examining how AI can enhance customer satisfaction, streamline service delivery, and create new value propositions, thereby reinforcing the importance of early adoption in gaining competitive advantages. This study contributes empirical insights to our understanding of innovation diffusion processes and their implications for entrepreneurship and technology adoption. By contextualizing the adoption of AI within the entrepreneurial mindset, the research provides a nuanced understanding of how innovation diffusion theories apply to contemporary technological advancements and their impact on startup success and customer experiences.

3 Literature Review

3.1 AI in Business and Innovation

Empowering Computers and Industry Applications

Artificial Intelligence, as conceptualized by Rich, emphasizes the goal of empowering computers to undertake tasks that are currently dominated by humans, thereby extending the capabilities of machines beyond their traditional functions. Today, AI has permeated various domains, demonstrating its versatility and transformative potential. This widespread adoption underscores the significant advancements in AI technologies and their capacity to revolutionize diverse sectors by enhancing efficiency, accuracy, and overall performance [8].

Optimizing Research and Innovation

Bole and others highlight strategies for optimizing research and innovation investments, emphasizing the crucial role of academia-industry collaboration. By fostering strong partnerships, these collaborations leverage the strengths of both academia's theoretical and experimental expertise and industry's practical and market-oriented focus [9]. Bruno, along with fellow researchers, explored the intricate relationship between Research and Development (R&D) activities and innovation performance within organizations. Their investigation sought to uncover the extent to which investments in R&D initiatives contribute to fostering innovation and driving organizational success [10].

Entrepreneurial Communication and Product Management

The potential of AI, particularly ChatGPT, in enhancing entrepreneurial communication has been observed in certain studies [11]. Particularly, modern platforms such as Bard, ChatGPT, the metaverse, and artificial intelligence offer innovative and immersive methods for practicing employer branding. The study done by Pandey presents a framework for organizations to align with advanced employer branding activities utilizing these cutting-edge tools and platforms [12]. Moreover, the pivotal role of Product Managers in start-ups was also underscored in another study based on systematic literature review which proposed a startup-specific framework to guide Product Manager activities towards maximizing expected value, thereby improving early-stage product decision-making and startup success rates [13].

AI in Customer Service and Business Automation

George and George discussed ChatGPT's applications in customer service. They furthered that the potential applications of ChatGPT are broad, including serving as a virtual assistant to boost operational efficiency in entrepreneurship. Its ability to comprehend natural language and offer tailored responses makes it versatile across various contexts. ChatGPT signifies a notable advancement in AI and natural language processing, with applications spanning customer service, language translation, education, and entertainment. Features like topic and emotion detection enable more authentic interactions between users and bots. The article suggests ChatGPT's potential in enhancing e-commerce, education, finance, news, and productivity sectors. While it promises to revolutionize technology interaction and improve daily life, challenges such as ethical use, biases, and user privacy must be addressed. Nonetheless, continued research and development may lead ChatGPT and similar AI technologies to define the next major era in history [14]. Montobbio and others emphasized two main areas regarding the rapid advancement of robotic technology, artificial intelligence, and automation: Firstly, there's a focus on analyzing tasks and occupations that are vulnerable to automation. Secondly, attention has been given to the labor market effects of industrial robots among those who adopt them [15]. Staccioli and Virgillito state that Labor-saving (LS) heuristics have historically influenced the pace and direction of technological advancement since the First Industrial Revolution. Automated systems, including robotic systems, are widely employed across various industries to cut labor costs and boost productivity. Compared to human operators, automated systems often offer cost advantages. Their implementation leads to enhanced information management, service provision, and data handling, resulting in increased efficiency, time savings, reduced manpower requirements, and significant cost reductions. LS patents are notably concentrated in fields such as transport, storage, packaging, diagnosis and therapy, digital information transmission, optical elements, and chemical or physical laboratory apparatus [16], and Yun and others analyzed the status and recent growth of technology and business models in the autonomous car and intelligent robot industries, using patents as indicators. They concluded three main points. Firstly, for industries in transition like autonomous cars, a business model-focused strategy is beneficial in the early stage, while a technology-focused strategy is effective in the mature stage. Secondly, in emerging industries like intelligent robots, a technology-focused strategy holds slightly more value in the early stage, but becomes more balanced with a business model-focused strategy in the mature stage. Thirdly, in mature stages of industries such as intelligent robots in Canada or autonomous cars in Japan, supplementing technology with a strong business model is crucial for continued success. [17] There are several other research studies that provide insights into Industry 4.0 [18–20].

3.2 Innovation Process

Perspectives on Innovation

In recent years, there has been notable advancement in AI Creativity, characterized by the generation of highly novel and suitable ideas, solutions, or outputs by autonomous machines. Amabile advocated for organizational researchers in creativity and innovation to focus considerable attention on studying artificial intelligence and computer-assisted human intelligence which includes exploring how these technologies can lead to creative breakthroughs and examining their potential impacts on and interactions with workers, consumers, organizations, and society at large [21]. A study by Kijkuit and van den Ende aimed to bridge the gap between idea generation and project selection, by introducing a dynamic framework that spans from idea generation to selection within the context of social networks. The framework suggests the necessity of strong ties and prior related knowledge, the involvement of decision makers, and a transition over time from large, non-redundant, and heterogeneous networks to smaller and more cohesive structures [22].

3.3 Innovation Process

Threats to Employment

Clark and colleagues explored the various impacts of technological advancements on employment [23]. It is observed that artificial intelligence is rapidly advancing to the point where many "good jobs" could become obsolete. Paralegals, journalists, office workers, and even computer programmers are at risk of being replaced by robots and smart software. This technological progress threatens both blue-collar and white-collar jobs, putting additional pressure on working- and middle-class families. Concurrently, these households face rising costs, particularly in education and healthcare, sectors that have not yet been significantly impacted by information technology. This situation could lead to widespread unemployment and inequality, potentially collapsing the consumer economy [24].

3.4 AI in Healthcare

Active Integration in Healthcare

Early investigations have demonstrated AI's active integration in healthcare settings, spanning diagnostics, therapy, healthcare management, and drug development [25–27]. Healthcare, characterized by its conservatism and high error costs, rigorously scrutinizes new advancements through regulatory oversight and medical expertise [28, 29].

Transformative Potential in Healthcare

Nevertheless, AI is increasingly recognized as a transformative force in healthcare, promising to improve access to quality healthcare, minimize errors, and enhance medical training [30, 31]. In healthcare, AI's primary function lies in accessing crucial information and facilitating decision-making processes [32]. Its superiority over human capabilities stems from its ability to process vast datasets [33].

Optimizing Business Processes

Beyond healthcare, AI enterprises are actively seeking to optimize business processes across diverse industries [34]. Technology not only streamlines existing healthcare practices but also offers avenues for innovation and value co-creation. Research within an Italian context leverages content analysis and data mining frameworks to illuminate technology's potential in healthcare. This involves utilizing software applications to enable on-demand access to medical products and prioritizing patient-centric services through E-commerce [35].

Telehealth and Digital Healthcare Innovations

The emergence of telehealth start-ups further accelerates innovation within the digital healthcare landscape. The efficient and affordable delivery of healthcare services is a critical goal for national governments, yet challenges persist in reaching everyone. Startups are attempting to address these gaps by introducing innovative solutions to reach underserved markets, although empirical evidence remains sparse. While there's evidence of technology adoption and acceptance of services based on effectiveness and affordability, challenges include licensing, policies, data privacy, security, and limited technology access among healthcare seekers. Research on the impact of health tech startups on healthcare service delivery remains nascent, with insufficient focus on entrepreneurship, business frameworks, and regulations [36].

The integration of AI technologies continues to reshape industries and drive innovation across various domains. From entrepreneurial communication to healthcare management, AI-driven innovations offer unprecedented opportunities for organizations to optimize processes, enhance customer experiences, and foster growth. The transformative potential of AI is evident in its ability to streamline existing practices, facilitate decision-making, and enable personalized interventions in healthcare. As startups and established enterprises alike embrace AI technologies, it is imperative to leverage collaborative frameworks, foster entrepreneurial mindsets, and prioritize user-centric design principles to maximize the benefits of AI-driven innovations.

3.5 *Ethics, AI and Healthcare*

Given the preface of health scenario in India where women are discriminated and due to systemic biases may not feel comfortable to discuss on sensitive subjects like sexual and reproductive health and rights, "Just Ask Chatbot" is a commendable initiative. It also boasts of using UNFPA's ethical guidelines in its design. However, majorly chatbots mimic human conversation by gathering data and identifying patterns to predict responses. Basic chatbots typically operate on predefined rules, using programmed decision trees or straightforward "If-Then-Else" logic [37]. Chatbots operate continuously without needing salaries or sick leave, though they still require human oversight for monitoring and updates. If they malfunction, they can be either upgraded or deactivated. These characteristics have increased interest from businesses and organizations, particularly in mental health [38]. Research suggests that several obstacles hinder the adoption of chatbots, including privacy concerns, financial limitations, skepticism, and a lower willingness among individuals from disadvantaged socioeconomic backgrounds [39]. Studies also reveal that adolescents face primary barriers such as stigma, embarrassment, low health literacy, and a preference for self-reliance [40]. For chatbots to be effective, it is crucial that users trust the technology to address their needs safely and efficiently, and that developers are attentive to user feedback and issues [41]. In AI healthcare applications, **inaccurate results** and **data leakage** are critical concerns. Inaccurate results can stem from flawed algorithms or biased data, leading to incorrect diagnoses and compromised patient safety. Data leakage involves the unintended exposure of sensitive patient information during AI model training, posing privacy risks [42, 43]. These issues highlight the need for rigorous validation, strong data security, and ethical oversight to ensure that AI technologies improve healthcare quality without compromising patient trust.

4 Research Gap

The literature reviewed sheds light on various aspects of entrepreneurship, technology, and innovation. However, amidst these discussions, there's a noticeable gap in research focusing on "The Impact of Perceived entrepreneurial mindset on Customer Experience in AI-Driven Startups." While studies explore the role of AI and entrepreneurship [10, 12], and discusses AI's potential in customer service [13], there is a lack of direct investigation into how the perceived entrepreneurial mindset in AI-driven startups influences customer experience. Understanding this relationship is crucial for startups leveraging AI technologies, as it could provide insights into optimizing customer interactions and fostering business growth. Therefore, our research aims to address this gap by examining how entrepreneurial attitudes, strategies, and decision-making processes within AI-driven startups impact the quality of customer experience and subsequent business outcomes particularly in health settings where there are several ethical concerns [39–43].

5 Methodology

The study is mixed in nature and employs both qualitative and quantitative methods. It is a correlational, cross-sectional study in which the data was collected from 58 users of Just Ask Chatbot—A unique Chatbot designed to play a significant role in empowering and supporting respondents in their journey towards better sexual and reproductive health. Purposive and snowball sampling techniques were used to select the participants within a month. This research aligns with the Diffusion of Innovation Theory by E.M. Rogers by framing its investigation within this theoretical framework. It explores how customer's perceived entrepreneurial mindset accelerates the adoption of AI innovations among startups, drawing parallels with the theory's emphasis on understanding adoption dynamics. By focusing on the traits associated with innovators and early adopters, the study examines how proactive startups that are more inclined to embrace AI technologies, drive their adoption within startup ecosystems resulting in customer acceptance. Partial Least Squares Structural Equation Modeling (PLS-SEM) and NVivo analysis were employed to assess the collected data using the software SmartPLS 4 and NVivo 14. Additionally, the research also investigates the consequences of AI adoption on customer experiences, extending the theory's focus on the outcomes of innovation diffusion. Overall, this study contributes empirical insights to our understanding of innovation diffusion processes and their implications for entrepreneurship and technology adoption.

5.1 Objectives

1. To assess the influence of perceived entrepreneurial mindset on the adoption of Artificial Intelligence (AI) technologies by customers within the context of Just Ask Chatbot.
2. To investigate the extent to which AI adoption affects the enhancement of customer experiences among users of Just Ask Chatbot.
3. To analyze the interplay between perceived entrepreneurial mindset, AI adoption, and customer experience enhancement, employing a mixed-method approach combining qualitative and quantitative techniques.
4. To identify key factors that facilitate or hinder the successful integration of AI-driven innovations, specifically focusing on Just Ask Chatbot, within the healthcare startup ecosystem.

5.2 Hypotheses

1. **Hypothesis1**

   Null Hypothesis (H0): There is no significant relationship between perceived entrepreneurial mindset and enhanced customer experience.

   Alternative Hypothesis (H1): Perceived entrepreneurial mindset is positively associated with enhanced customer experience.
2. **Hypothesis 2**

   Null Hypothesis (H0): There is no significant relationship between perceived entrepreneurial mindset and innovation adoption through AI.

   Alternative Hypothesis (H1): Perceived entrepreneurial mindset positively influences innovation adoption through AI.
3. **Hypothesis 3**

   Null Hypothesis (H0): There is no significant relationship between innovation adoption through AI and enhanced customer experience.

   Alternative Hypothesis (H1): Innovation adoption through AI positively contributes to enhanced customer experience.

6 Data Analysis and Findings

6.1 Construct Reliability and Validity

Table 1 examines the construct's reliability and validity. The reliability of the measures in this study was assessed through several indicators. Cronbach's alpha values, which gauge internal consistency reliability, were found to range from 0.703 to 0.865. These values fall within the acceptable to excellent range, indicating a satisfactory level of consistency in the responses across items within each construct. Additionally, composite reliability (rho_a) values ranged from 0.708 to 0.879, indicating good to excellent reliability across the constructs. Another measure of reliability, composite reliability (rho_c), ranged from 0.87 to 0.917, demonstrating good to very high-reliability levels. These findings collectively suggest that the measures used in the study exhibit robust reliability, enhancing confidence in the validity of the research outcomes.

The study also examined the convergent validity of the measures using the Average Variance Extracted (AVE) values. The AVE values ranged from 0.77 to 0.795, indicating satisfactory levels of convergent validity across the constructs. These values suggest that between 77 and 79.5% of the variance in the constructs is captured by their respective indicators. In other words, the indicators within each construct share a substantial amount of variance, supporting the notion that they are measuring the same underlying construct. This finding adds further confidence to the validity of the measurement model used in the study.

Table 2 examines the Heterotrait-Monotrait (HTMT) ratio, and reveals that the values for each pair of constructs are below the recommended threshold of 0.85, indicating that the constructs are distinct from each other. Specifically, the HTMT values between Enhanced Customer Experience (ECE) and Perceived entrepreneurial mindset (EM), as well as between ECE and Innovation Adoption through AI (IA), are below 0.85. Similarly, the HTMT values between EM and IA are also below 0.85.

Table 1 Construct reliability

	Cronbach's alpha	Composite reliability (rho_a)	Composite reliability (rho_c)	Average variance extracted (AVE)
Enhanced customer experience	0.742	0.742	0.886	0.795
Perceived entrepreneurial mindset	0.703	0.708	0.87	0.77
Innovation adoption through AI	0.865	0.879	0.917	0.787

Table 2 Discriminant validity HTMT ratio

	Enhanced customer experience	Perceived entrepreneurial mindset	Innovation adoption through AI
Enhanced customer experience			
Perceived entrepreneurial mindset	0.84		
Innovation adoption through AI	0.249	0.492	

Table 3 Discriminant validity Forner Larker

	Enhanced customer experience	Perceived entrepreneurial mindset	Innovation adoption through AI
Enhanced customer experience	0.891		
Perceived entrepreneurial mindset	0.679	0.878	
Innovation adoption through AI	0.2	0.391	0.887

The Fornell-Larcker criterion demonstrated in Table 3 shows that the square roots of the Average Variance Extracted (AVE) for each construct are higher than the correlations between the constructs, indicating discriminant validity.

6.2 *Structured Equation Modelling*

The cross-loading results shown in Table 4 indicate that each item loads more strongly on its respective construct compared to others, further supporting the discriminant validity of the model. Thus the model demonstrates adequate discriminant validity, indicating that the constructs in the study are distinct from each other.

Figure 2 depicts PLS SEM model—Structural Model depicting the impact of innovation adoption and perceived entrepreneurial mindset on customer experience while Fig. 3 depicts PLS SEM bootstrapping depicting the impact of innovation adoption and perceived entrepreneurial mindset on customer experience. To improve the assessment of the indirect impacts of the independent variable (Perceived entrepreneurial mindset) on the dependent variable (Enhanced customer experience) via mediating variable (Innovation adoption through AI), the bootstrapping method was employed with bootstrap subsamples of 1000.

Table 4 Cross loadings

	Enhanced customer experience	Perceived entrepreneurial mindset	Innovation adoption through AI
ECE1	0.891	0.599	0.195
ECE2	0.892	0.611	0.161
EM3	0.603	0.891	0.4
EM4	0.589	0.864	0.281
IA1	0.203	0.377	0.897
IA2	0.147	0.373	0.918
IA4	0.181	0.281	0.845

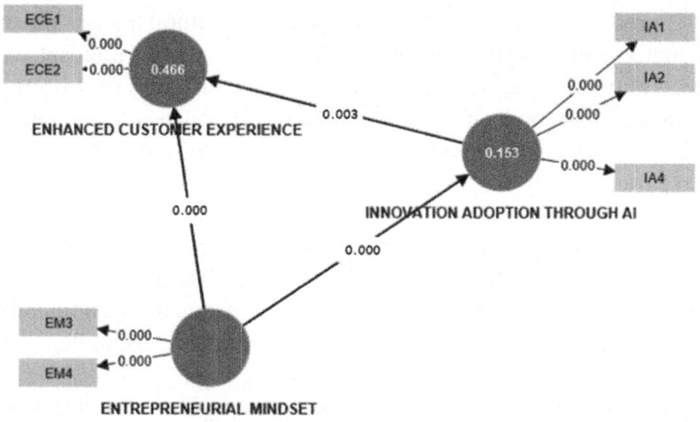

Fig. 2 PLS SEM model—Structural Model depicting the impact of innovation adoption and perceived entrepreneurial mindset on customer experience

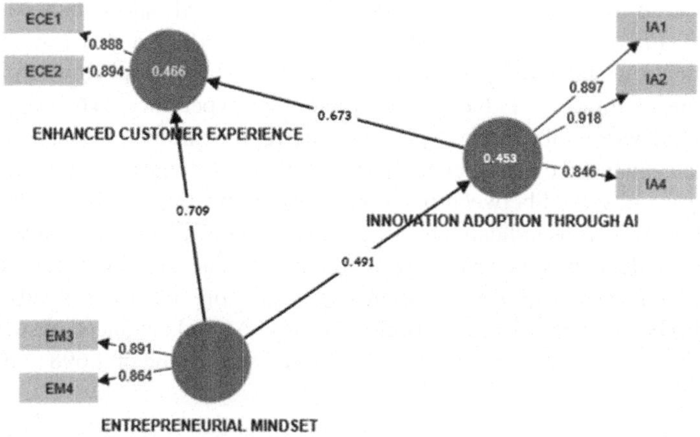

Fig. 3 PLS SEM bootstrapping model—Structural Model depicting the impact of innovation adoption and perceived entrepreneurial mindset on customer experience

Utilizing a structural equation model (SEM) to investigate the interrelationships among three primary constructs: "Enhanced Customer Experience," "Innovation Adoption Through AI," and "Entrepreneurial Mindset," it was found that "Entrepreneurial Mindset" significantly influences both "Enhanced Customer Experience" and "Innovation Adoption Through AI," with path coefficients of 0.000 indicating strong, direct effects. The variance explained (R2) for "Enhanced Customer Experience" is 0.466, suggesting that nearly half of the variance is accounted for by the model, while the R2 for "Innovation Adoption Through AI" is 0.153, indicating a more modest but still significant influence. These findings highlight the pivotal role of an entrepreneurial mindset in driving customer experience enhancements and facilitating AI-driven innovation adoption, underscoring its importance in organizational strategies.

Figure 2 presents a PLS-SEM bootstrapping structural model illustrating the impact of innovation adoption and perceived entrepreneurial mindset on customer experience. The model indicates that the entrepreneurial mindset exerts a significant direct influence on both enhanced customer experience and innovation adoption through AI, with path coefficients of 0.000, denoting strong, direct effects. Enhanced customer experience is significantly predicted by entrepreneurial mindset, explaining 46.6% of its variance ($R^2 = 0.466$). Meanwhile, innovation adoption through AI is influenced by both enhanced customer experience and entrepreneurial mindset, with a combined explained variance of 15.3% ($R^2 = 0.153$). These findings underscore the critical role of fostering an entrepreneurial mindset to enhance customer experiences and facilitate the adoption of AI innovations, highlighting its strategic importance for organizations.

With Table 5 it is observed that perceived entrepreneurial mindset and innovation adoption through artificial intelligence result in 46.6% of change in enhanced customer experiences.

Table 6 reveal compelling insights into the relationships among perceived entrepreneurial mindset, innovation adoption through AI, and enhanced customer experience. Firstly, the strong T statistics value of 8.923 and a low *P* value of 0.001 indicate a statistically significant positive relationship between perceived entrepreneurial mindset and enhanced customer experience. This suggests that entrepreneurs who exhibit a strong perceived entrepreneurial mindset are more likely to deliver an enhanced experience to their customers. Secondly, a significant positive relationship is observed between perceived entrepreneurial mindset and innovation adoption through AI, as indicated by a T statistics value of 4.067 and a low *P* value of 0.001. This underscores the role of perceived entrepreneurial mindset in driving the adoption of innovative technologies such as AI within organizations. Lastly, while the relationship between innovation adoption through AI and enhanced customer experience is also statistically significant, with a T statistics value of 2.698 and a *P* value

Table 5 R^2

	R square	R square adjusted
ECE	0.466	0.445

Table 6 Path coefficients

	Original sample (O)	Sample mean (M)	Standard deviation (STDEV)	T statistics (|O/STDEV|)	*P* values
Perceived entrepreneurial mindset -> Enhanced customer experience	0.709	0.714	0.08	8.923	0.001
Perceived entrepreneurial mindset -> Innovation adoption through AI	0.391	0.403	0.096	4.067	0.001
Innovation adoption through AI -> Enhanced customer experience	0.221	0.226	0.082	2.698	0.003

of 0.003, the effect size appears slightly weaker compared to the other relationships. Nonetheless, this finding suggests that organizations embracing AI technologies are likely to see improvements in their customer experience, albeit to a slightly lesser extent compared to the direct influence of perceived entrepreneurial mindset. Overall, these results highlight the importance of both perceived entrepreneurial mindset and innovation adoption through AI in enhancing customer experience, offering valuable insights for businesses aiming to leverage these factors to achieve competitive advantage.

6.3 Model Fit

The model fit indices indicate that both the estimated model and the saturated model demonstrate an acceptable fit to the data. The Standardized Root Mean Square Residual (SRMR) for both models is 0.08, suggesting a good fit. Additionally, the discrepancy measures (d_ULS and d_G) for both models are identical at 0.177, indicating consistency between the estimated and saturated models. The Chi-square values for both models are also equivalent at 61.549. However, the Normed Fit Index (NFI) for both models is relatively low at 0.67, indicating a moderate fit to the data. Overall, while the NFI suggests some limitations in model fit, the other fit indices, including SRMR and discrepancy measures, support the adequacy of both the estimated and saturated models for the study (Table 7).

Table 7 Model Fit

	Saturated model	Estimated model
SRMR	0.08	0.08
d_ULS	0.177	0.177
d_G	0.16	0.16
Chi-square	61.549	61.549
NFI	0.67	0.67

6.4 Content Analysis

Figure 4 depicts a word cloud which centers around themes related to customer interactions, support, data, and business innovation. The largest word, "chatbots," indicates that this is the focal point of the concept. Surrounding it are words like "respondents," "information," "appreciated," and "support" which suggest that the word cloud is related to customer service or customer perceptions, in the context of AI or chatbots like Just Ask Chatbot. Words like "reproductive," "sexual", "empowerment", "analysis", "accessibility," "reliable," and "information" imply a focus on customer interaction with the chatbot. The above word cloud represents the features or benefits of a customer relationship management system, an AI chatbot service that focuses on customer data. It captures a broad range of activities and objectives that are important for businesses seeking to enhance their customer interaction strategies using technology.

It was found that the customers were not much aware about the privacy of their data stored. However, they were confident on the delete option which is provided in the chatbot. Moreover, they shared that the bot did not ask much of the personal details, just an inquiry into their city/town/village of residence, their gender and age.

Fig. 4 Word cloud to gauge central themes of the interview

Further, there are pre-defined set of options to which the responses are provided by the chatbot.

Figure 5 presents a graphical representation of sentiment analysis results to understand how the customers perceive integration of chatbot in health services. The largest section of the graph is labeled "Positive," which dominates the visual space. This suggests that the sentiment towards the integration of AI in health services is overwhelmingly positive. The prominence of the positive section indicates that participants generally see several advantages to this integration. There is a much smaller section labeled "Neutral," which is significantly less prominent than the positive section. This indicates that a smaller proportion of the sentiments expressed were neither positive nor negative, suggesting that some people see the integration as neither beneficial nor detrimental, or they might lack enough information to form an opinion. There is no visible section labeled "Negative," which implies that there were no negative sentiments or not enough to warrant a section that expressed a clear disadvantage or criticism about integrating cohabits in health services. Therefore the interpretation of the image would be that the sentiment towards integrating chatbots with health services is largely positive, with few neutral positions and seemingly no negative sentiments recorded. It suggests that the consensus is that there are several advantages to the integration of this technology into health services like Just Ask Chatbot.

6.5 Ethical Concerns in Using Just Ask Chatbot

Although it is too early to comment upon the ethical issues pertaining to the young chatbot - Just Ask Chatbot, as it is recently launched in India, still, the participants were asked about their ethical concerns while using Just Ask Chatbot. Figure 6 represents a word cloud based on their responses.

The word cloud in Fig. 6 reveals key themes emerging from participant interviews regarding ethical concerns in using the Just Ask chatbot. Users appear to be primarily focused on data privacy and security, as evidenced by the prominence of words like "data," "privacy," "security," and "confidentiality." There's a strong emphasis on the ethical handling of information, with participants expressing concerns about how their personal data is stored, processed, and protected. The word "trust" features prominently, suggesting users are grappling with issues of reliability and trustworthiness in their interactions with the chatbot. Technical aspects and the accuracy of responses also seem to be significant concerns. Participants appear to be seeking reassurance about the chatbot's ability to handle sensitive information and complex queries ethically. The presence of words like "framework," "standards," and "measures" indicates a desire for clear guidelines and safeguards in the chatbot's operation. Overall, the word cloud paints a picture of users who are cautiously engaging with the technology while remaining acutely aware of potential risks and ethical implications.

Fig. 5 Sentiment analysis

Fig. 6 Word cloud to understand ethical concerns in Just Ask Chatbot

6.6 Key Findings

Perceived Entrepreneurial Mindset and AI Adoption

The study found a significant positive relationship between perceived entrepreneurial mindset and innovation adoption through AI among users of Just Ask Chatbot. Participants who exhibited a strong perceived entrepreneurial mindset were more inclined to embrace AI technologies, indicating the importance of entrepreneurial attitudes in driving technological innovation adoption within healthcare startup ecosystems.

AI Adoption and Customer Experience Enhancement

The research revealed a statistically significant positive relationship between innovation adoption through AI and enhanced customer experience among Just Ask Chatbot users. This suggests that organizations leveraging AI technologies, such as Just Ask Chatbot, are likely to witness improvements in customer experiences, highlighting the potential of AI-driven innovations to enhance user satisfaction and engagement.

Interplay between Perceived Entrepreneurial Mindset, AI Adoption, and Customer Experience Enhancement

By analyzing the interplay between perceived entrepreneurial mindset, AI adoption, and customer experience enhancement, the study identified a complex relationship where perceived entrepreneurial mindset positively influences both AI adoption and customer experience enhancement. This underscores the critical role of entrepreneurial attitudes and strategic decision-making processes in driving the successful integration of AI-driven innovations within healthcare startup ecosystems, ultimately leading to enhanced customer experiences.

Identification of Key Facilitating Factors and Barriers

Through qualitative analysis and interviews, key facilitating factors and barriers to the successful integration of AI-driven innovations, specifically Just Ask Chatbot, within healthcare startup ecosystems were identified. Facilitating factors included proactive entrepreneurial attitudes, organizational culture conducive to innovation, and user-centric design principles, while barriers encompassed technological complexity, regulatory constraints, and resistance to change.

Further, the users' main concerns with the Just Ask chatbot were identified as data privacy, security, trust, technical accuracy and lack of confidence regarding handling of information.

7 Conclusion

In this chapter, the authors examined the intricate relationship between perceived entrepreneurial mindset, AI adoption, and its repercussions on customer experiences among 58 individuals using Just Ask Chatbot i.e., a startup environment particularly focusing on healthcare domains. Drawing from E.M. Rogers' Diffusion of Innovation Theory, we explored how entrepreneurial traits shape the integration of AI technologies and subsequent customer engagement and loyalty. Our analysis uncovered several key insights: firstly, a significant positive correlation exists between perceived entrepreneurial mindset and enhanced customer experience, underscoring the pivotal role of entrepreneurial attitudes in shaping customer interactions. Secondly, perceived entrepreneurial mindset positively influences innovation adoption through AI, highlighting the inclination of proactive entrepreneurs towards embracing AI technologies within startup ecosystems. Lastly, innovation adoption through AI contributes to enhanced customer experience, emphasizing the role of AI in improving customer satisfaction. These findings offer practical implications for startup owners and industry stakeholders, advocating for environments conducive to entrepreneurial experimentation and strategic AI adoption to drive sustainable growth and competitive advantage in the digital era. Overall, our research advances understanding of innovation diffusion processes and their impact on entrepreneurship and technology adoption, providing valuable insights for businesses navigating the complexities of the modern digital landscape. Integrating AI in healthcare requires a strong ethical framework, as promised by the Just Ask Chatbot. Though no instances are recorded, participants revealed concerns about privacy, security, trust, and accuracy. It is way too early to assess the UNFPA's guidelines' success in Just Ask Chatbot.

References

1. Rogers, E.M.: Diffusion of Innovations, 5th edn. Free Press, New York (2003)
2. Saikia, N., Moradhvaj, Bora, J.K.: Gender difference in health-care expenditure: evidence from India human development survey. PloS one, **11**(7) (2016). https://doi.org/10.1371/journal.pone.0158332
3. Moradhvaj, S.N.: Gender disparities in health care expenditures and financing strategies (HCFS) for inpatient care in India. SSM Popul. Health **9**, 100372 (2019). https://doi.org/10.1016/j.ssmph.2019.100372
4. Singh, S.: Gender disparity in education and health in India: a comprehensive analysis. Int. J. English Lit. Soc. Sci. **7**(1), 202–210 (2022)
5. Mohanan, M., Hay, K., Mor, N.: Quality of health care in India: challenges, priorities, and the road ahead. Health Aff. **35**(10), 1753–1758 (2016). https://doi.org/10.1377/hlthaff.2016.0676
6. Wojnar, A.M.: Women's reproductive autonomy as the new watchword. UNFPA India (2023). https://india.unfpa.org/en/news/womens-reproductive-autonomy-new-catchword
7. UNFPA: In India, UNFPA launches 'Just Ask!'—A chatbot for sexual and reproductive health and rights awareness. UNFPA (2023). https://www.unfpa.org/updates/india-unfpa-launches-just-ask-chatbot-sexual-and-reproductive-health-and-rights
8. Rich, E.: Artificial intelligence and the humanities. Comput. Humanit. **19**(2), 117–122 (1985)
9. Bole, D., Galabova, L., Haley, C., Kokorotsikos, P., Matanovac-Vučković, R., Rizzuto, C., Taylor, S., Vladut, G., Zambelli, M.: Strategic evaluation of the technology transfer and IPR protection systems of Bulgaria, Croatia and Romania and recommendations for their enhancement. Publications Office of the European Union, Belgium (2024). https://doi.org/10.2760/375746, JRC136807

10. Bruno, R.L., Crescenzi, R., Estrin, S., Petralia, S.: Multinationals, innovation, and institutional context: IPR protection and distance effects. J. Int. Bus. Stud. **53**(9), 1945–1970 (2021). https://doi.org/10.1057/s41267-021-00452-z
11. Short, C., Short, J.: The artificially intelligent entrepreneur: ChatGPT, prompt engineering, and entrepreneurial rhetoric creation. J. Bus. Ventur. Insights **19**, e00388 (2023). https://doi.org/10.1016/j.jbvi.2023.e00388
12. Pandey, N.: Future of employer branding in the era of BarD, ChatGPT, Metaverse and artificial intelligence. NHRD Netw. J. **16**(3), 258–268 (2023). https://doi.org/10.1177/26314541231170434
13. Pattyn, F.: Preliminary structured literature review results using ChatGPT: towards a pragmatic framework for product managers at software startups. In: IEEE 31st International Requirements Engineering Conference (RE). IEEE (2023)
14. George, A.S., George, A.S.H.: A review of ChatGPT AI's impact on several business sectors. Partners Universal Int. Innov. J. **1**(1), 9–23 (2023). https://doi.org/10.5281/zenodo.7644359
15. Montobbio, F., Staccioli, J., Virgillito, M.E., Vivarelli, M.: Robots and the origin of their labour-saving impact. Technol. Forecast. Soc. Chang. **174**, 121122 (2022). https://doi.org/10.1016/j.techfore.2021.121122
16. Staccioli, J., Virgillito, M.E.: Back to the past: the historical roots of labor-saving automation. Eurasian Bus. Rev. **11**, 27–57 (2021). https://doi.org/10.1007/s40821-020-00179-1
17. Yun, J.J., Won, D., Park, K., Jeong, E., Zhao, X.: The role of a business model in market growth: The difference between the converted industry and the emerging industry. Technol. Forecast. Soc. Change. **146**, 534–562 (2019). https://doi.org/10.1016/j.techfore.2019.04.024
18. Cetrulo, A., Nuvolari, A.: Industry 4.0: revolution or hype? Reassessing recent technological trends and their impact on labour. J. Ind. Bus. Econ. **46**(3), 391–402 (2019). https://doi.org/10.1007/s40812-019-00132-y
19. Pardi, T., Krzywdzinski, M., Lüthje, B.: Digital manufacturing revolutions as political projects and hypes: evidences from the auto sector. In: ILO Working Paper 3, Apr 2020. https://labordoc.ilo.org/permalink/41ILO_INST/j3q9on/alma995073193302676
20. Cirillo, V., Rinaldini, M., Staccioli, J., Virgillito, M.E.: Technology versus workers: the case of Italy's Industry 4.0 factories. Struct. Change Econ. Dyn. **56**, 166–183 (2021). https://doi.org/10.1016/j.strueco.2020.09.007
21. Amabile, T.M.: Guidepost: creativity, artificial intelligence, and a world of surprises GuidePost letter for academy of management discoveries. Academy of Management Discoveries (2019). https://doi.org/10.5465/amd.2019.0075
22. Kijkuit, B., Van Den Ende, J.: The organizational life of an idea: integrating social network, creativity and decision-making perspectives. J. Manage. Stud. **44**(6), 863–882 (2007). https://doi.org/10.1111/j.1467-6486.2007.00695.x
23. Clark, J., Freeman, C., Soete, L.: Long waves and technological developments in the 20th century. Konjunktur Krise Gesellschaft **25**(2), 132–169 (1981)
24. Ford, M.: The Rise of the Robots: Technology and the Threat of Mass Unemployment. Basic Books (2015)
25. Ivan, M., Velicanu, M.: Healthcare industry improvement with business intelligence. Informatica Economica **20**(2), 81–89 (2015)
26. Fleming, N.: How artificial intelligence is changing drug discovery. Nature **557**(7707), S55–S57 (2018)
27. Hamet, P., Tremblay, J.: Artificial intelligence in medicine. Metabolism **69**, S36–S40 (2017)
28. Jiang, F., Jiang, Y., Zhi, H., Shen, H., Wang, Y.: Artificial intelligence in healthcare: past, present and future. Stroke Vasc. Neurol. **2**(4), 230–243 (2017)
29. Kulkov, I.: The role of artificial intelligence in business transformation: a case of pharmaceutical companies. Technol. Soc. **66**, 101629 (2021)
30. Mak, K., Pichika, M.R.: Artificial intelligence in drug development: present status and future prospects. Drug Discovery Today **24**(3), 773–780 (2019)
31. Miller, D., Brown, E.: Artificial intelligence in medical practice: the question to the answer? Am. J. Med. **131**(2), 129–133 (2018)

32. Murdoch, T., Detsky, A.: The inevitable application of big data to health care. J. Am. Med. Assoc. **309**(13), 1351–1352 (2013)
33. Buch, V., Ahmed, I., Maruthappu, M.: Artificial intelligence in medicine: current trends and future possibilities. Br. J. Gen. Practice: J. Roy. Coll. Gen. Pract. **689**(668), 143–144 (2018)
34. Brynjolfsson, E., McAfee, A.: The business of artificial intelligence. Harv. Bus. Rev. **7**, 3–11 (2017)
35. Morande, S., Del Vacchio, E., Ranieri, A.: Digital innovations in healthcare startups: transforming service ecosystem. J. Bus. Manage. Stud. **2**(1), 26–39 (2020). Retrieved from https://www.al-kindipublisher.com/index.php/jbms/article/view/38
36. Chakraborty, I., Ilavarasan, P.V., Edirippulige, S.: Health-tech startups in healthcare service delivery: a scoping review. Soc Sci Med **278**, 113949 (2021)
37. Lewis, S.: Ultimate guide to chatbots 2020—Examples, best practices & more (2020). https://blog.appsumo.com/ultimate-guide-to-chatbots-2020/
38. Sheridan Rains, L., Johnson, S., Barnett, P., Steare, T., Needle, J.J., Carr, S., Lever Taylor, B., Bentivegna, F., Edbrooke-Childs, J., Scott, H.R., Rees, J., Shah, P., Lomani, J., Chipp, B., Barber, N., Dedat, Z., Oram, S., Morant, N., Simpson, A., COVID-19 Mental Health Policy Research Unit Group: Early impacts of the COVID-19 pandemic on mental health care and on people with mental health conditions: framework synthesis of international experiences and responses. Soc. Psychiatry Psychiatric Epidemiol. **56**(1), 13–24 (2021). https://doi.org/10.1007/s00127-020-01924-7
39. Han, X.: Am I asking it properly?: designing and evaluating interview chatbots to improve elicitation in an ethical way. In: Proceedings of the 25th International Conference on Intelligent User Interfaces Companion, pp. 33–34. ACM, Cagliari, Italy (2020)
40. Crutzen, R., Bosma, H., Havas, J., et al.: What can we learn from a failed trial: insight into non-participation in a chat-based intervention trial for adolescents with psychosocial problems. BMC. Res. Notes **7**, 824 (2014)
41. Egger, F.N.: 'Trust me, I'm an online vendor': towards a model of trust for e-commerce system design. In: CHI '00 Extended Abstracts on Human Factors in Computing Systems, pp. 101–102. Association for Computing Machinery, New York, NY, USA (2000)
42. Neill, D.B.: Using artificial intelligence to improve hospital inpatient care. IEEE Intell. Syst. **28**, 92–95 (2013). https://doi.org/10.1109/MIS.2013.51
43. Gama, F., Tyskbo, D., Nygren, J., Barlow, J., Reed, J., Svedberg, P.: Implementation frameworks for artificial intelligence translation into health care practice: scoping review. J. Med. Internet Res. **24**(1), e32215 (2022). https://doi.org/10.2196/32215

Shumayela Hasan is a senior faculty member in Economics at The Bal Bhawan School. Formerly an Assistant Professor at The Bhopal School of Social Sciences, she brings over 18 years of teaching experience. Dr. Hasan has also served as Project Director for a research initiative funded by the Indian Council of Social Science Research. With multiple degrees in Economics, Finance, and Education, she is currently pursuing an MBA in Marketing. An accomplished author, she has published eight books and numerous articles in national and international journals.

Richi Simon is an Assistant Professor of Social Work at The Bhopal School of Social Sciences, with over 12 years of experience in teaching and research. She earned her PhD in Social Work from the Indore School of Social Work, focusing on Quality Education in Higher Education Institutions (HEIs) of Social Work. Her interdisciplinary academic background includes Social Work, Human Resource Management, and Computer Applications. She is currently pursuing MA Psychology. Dr. Simon is a dedicated researcher, with several published books, chapters, and research articles. She is a frequent presenter at national and international academic forums.

Ethical Considerations and Bias in AI Applications in Healthcare

Azamat Ali

Abstract This chapter examines the ethical issues and biases inherent in the application of artificial intelligence (AI) within healthcare, highlighting both the opportunities and challenges AI presents for enhancing patient care. Key ethical concerns include protecting patient privacy, obtaining informed consent, ensuring transparency, and establishing accountability within AI-driven decision-making processes. Biases within AI algorithms—arising from unbalanced data, flaws in algorithm design, or unintended human influences—pose significant challenges, potentially leading to disparities in healthcare delivery and exacerbating existing inequalities. To address these issues, this chapter proposes a comprehensive, patient-centered strategy focused on fairness and equal treatment. This approach emphasizes rigorous data management practices, thorough evaluations of algorithm performance, and sustained ethical oversight, while advocating for diverse representation in AI development teams and the involvement of all stakeholders in decision-making. By proactively addressing the ethical considerations and biases in AI implementation, healthcare providers can leverage AI's benefits while promoting equitable and ethical patient care.

Keywords Artificial intelligence (AI) · Healthcare ethics · Patient privacy · Informed consent · Algorithmic bias · Healthcare inequality · Transparency · Accountability · Fairness in AI · Patient-centered care

1 Introduction

The utilization of artificial intelligence in the healthcare sector is seeing tremendous growth, propelled by technological breakthroughs, the greater accessibility of healthcare data, and the rising acknowledgment of the potential advantages of AI in enhancing patient care. AI applications in healthcare encompass a wide

A. Ali (✉)
School of Liberal and Creative Arts, Lovely Professional University, Phagwara, Punjab, India
e-mail: azamatcug@gmail.com

N. R. Vajjhala et al. (eds.), *Artificial Intelligence in Healthcare Information Systems—Security and Privacy Challenges*, Information Systems Engineering and Management 34, https://doi.org/10.1007/978-3-031-84404-1_7

range of areas, including medical imaging, diagnostics, treatment optimization, drug discovery, predictive analytics, and personalized medicine [1]. AI has the potential to increase patient outcomes, improve efficiency, and decrease healthcare costs by enabling early illness identification and individualized treatment strategies [2].

AI algorithms in medical imaging may examine radiological images, including X-rays, MRIs, and CT scans, to identify anomalies, support diagnosis, and facilitate treatment planning [3]. These algorithms can help radiologists identify subtle abnormalities and prioritize urgent cases, leading to faster and more accurate diagnoses. In diagnostics, AI-powered tools can analyze clinical data, laboratory results, and patient histories to assist healthcare professionals in identifying diseases, predicting patient outcomes, and recommending appropriate treatment options [4]. Artificial intelligence algorithms can analyze large volumes of data in order to recognize patterns, trends, and connections that may not be easily noticeable to human physicians. This can result in the early identification and improved treatment of illnesses.

In drug discovery and development, AI technologies are being used to accelerate the identification of potential drug candidates, optimize drug formulations, and predict drug responses in specific patient populations. Through the analysis of chemical structures, biological processes, and clinical trial data, artificial intelligence (AI) algorithms can find potential therapeutic targets, create new compounds, and enhance the effectiveness and safety existing drugs [5]. AI-powered decision support systems can aid healthcare practitioners in optimizing therapy by helping them choose the most efficacious treatment choices for individual patients, considering their distinct features, medical histories, and genetic profiles. These systems have the capability to examine patient data, medical literature, and treatment standards to offer customized therapy suggestions that are specifically designed to meet the individual requirements and preferences of each patient.

Predictive analytics use AI algorithms to examine electronic health records, data from wearable devices, and other health information sources to detect individuals who are susceptible to acquiring specific diseases or consequences [6]. Through the early identification of high-risk patients, healthcare practitioners may take proactive measures to prevent negative outcomes, enhance patient results, and decrease healthcare expenses. AI technologies are now employed in personalized medicine to examine genomic data, biomarkers, and other patient-specific information. This analysis is used to customize treatment plans and treatments based on individual patients' genetic composition, lifestyle circumstances, and preferences. AI-powered solutions can enhance treatment results, reduce unwanted effects, and enhance patient satisfaction by offering tailored treatment suggestions [7].

However, alongside these advancements come complex ethical considerations and biases inherent in AI implementation in healthcare. As AI algorithms increasingly influence medical decision-making processes, there is a pressing need to address ethical issues to ensure patient safety, privacy, and equitable access to care. Moreover, biases within AI systems pose significant challenges, potentially leading to disparities in healthcare delivery and exacerbating existing inequalities in health outcomes [8]. Therefore, it is crucial to examine and address these ethical concerns and biases to

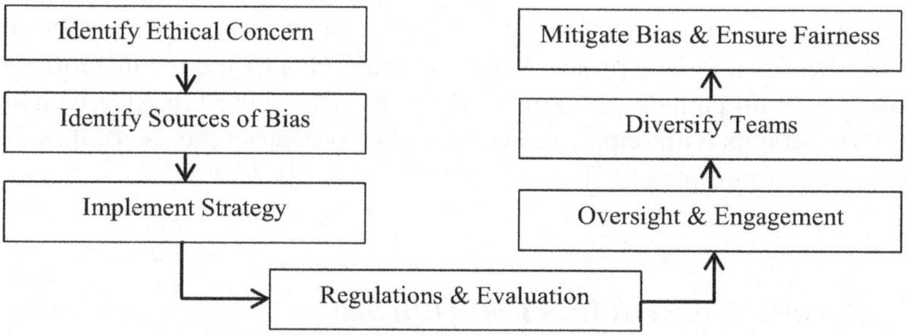

Fig. 1 Ethical considerations and bias in healthcare applications

maximize the benefits of AI technology while minimizing potential harm. Figure 1 addresses ethical considerations and bias in AI healthcare application.

2 Ethical Considerations in AI Applications in Healthcare

AI integration in healthcare has great potential to enhance patient care, optimize treatment results, and improve clinical decision-making in the fast-changing healthcare industry. Nevertheless, the use of AI in healthcare brings up noteworthy ethical concerns that need thorough attention to protect patient rights, privacy, and overall welfare. This section explores key ethical considerations in AI applications in healthcare, focusing on the importance of safeguarding patient privacy, obtaining informed consent, ensuring transparency in AI algorithms, and establishing accountability for AI-driven decisions [9].

2.1 Protecting Patient Privacy in AI Healthcare

AI implementation in healthcare frequently entails the acquisition, retention, and examination of significant quantities of confidential patient data, including medical records, diagnostic pictures, genetic information, and biometric data. Therefore, ensuring the protection of patient privacy is of utmost importance in order to preserve patient confidentiality, autonomy, and confidence in the healthcare system [10]. An important ethical concern in the use of AI in healthcare is the responsible handling of patient data, in compliance with established privacy principles and regulations, such as the “Health Insurance Portability and Accountability Act (HIPAA)” in the United States or the “General Data Protection Regulation (GDPR)” in the European Union. Healthcare companies and AI developers should use strong data security measures, encryption methods, and access restrictions to thwart unwanted access, disclosure, or misuse of patient data [11]. Moreover, healthcare providers must ensure that patient data is anonymized or de-identified whenever possible to minimize the risk of re-identification and protect patient privacy. Healthcare organizations should be transparent with patients about how their data will be used in AI-driven systems and provide patients with opportunities to opt-out or restrict the use of their data for research or other purposes [12].

2.2 Informed Consent in AI Healthcare

In healthcare, informed consent is a crucial ethical concept that mandates healthcare practitioners to seek voluntary and well-informed agreement from patients before performing any medical operations, treatments, or interventions. In the context of AI applications in healthcare, obtaining informed consent is essential to respect patient autonomy, promote transparency, and ensure that patients understand the potential risks and benefits of AI-driven interventions [13]. Healthcare professionals are obligated to communicate with patients regarding the utilization of artificial intelligence (AI) technology in their medical treatment. This includes providing information on how AI algorithms will evaluate their data, make predictions, and aid in clinical decision-making. Patients should get comprehensive and comprehensible information on the objectives of AI-driven treatments, the categories of data that will be gathered and processed, the potential consequences for their treatment, and any risks or constraints linked to AI technology [14]. Moreover, it is important that patients are given the chance to inquire, voice their worries, and make well-informed choices regarding their consent to the utilization of AI in their medical treatment. Healthcare practitioners must get express consent from patients before utilizing their data in AI-driven systems and must uphold patients' ability to revoke consent at any given moment [15].

2.3 AI Transparency in Decision-Making

Transparency is a core ethical principle in AI applications in healthcare, as it promotes accountability, trust, and understanding of how AI algorithms work, how decisions are made, and what factors influence the outcomes of AI-driven interventions. Transparent AI algorithms enable healthcare professionals and patients to evaluate the reliability, accuracy, and potential biases of AI predictions and recommendations, thereby enabling more informed decision-making and ensuring patient safety [16]. Healthcare organizations and AI developers must strive to enhance transparency in AI algorithms and decision-making processes by providing clear and accessible information about the design, development, and validation of AI models [17]. This includes disclosing the data sources used to train AI algorithms, the methodologies employed in algorithm development, the performance metrics used to evaluate algorithm accuracy and reliability, and any limitations or biases inherent in the AI system.

2.4 Accountability in AI Healthcare Decisions

Establishing accountability is essential in AI applications in healthcare to ensure that healthcare professionals, AI developers, and healthcare organizations are held responsible for the outcomes of AI-driven decisions and interventions. Accountability mechanisms help to identify errors, biases, or adverse effects of AI technologies, address patient concerns, and mitigate the risk of harm to patients [18]. Healthcare organizations should establish clear lines of accountability and responsibility for AI-driven decisions, ensuring that healthcare professionals are ultimately responsible for interpreting AI predictions, making clinical judgments, and determining the appropriate course of action for patient care. Moreover, healthcare organizations should provide training and support to healthcare professionals to enable them to understand the limitations and potential biases of AI algorithms and to use AI technologies effectively and responsibly in clinical practice [19]. AI developers have the obligation to oversee the design, development, and implementation of AI algorithms. They must ensure that AI systems undergo thorough testing, validation, and monitoring to guarantee optimal performance, accuracy, and fairness. Developers must incorporate tools to continuously monitor and evaluate AI algorithms to detect and rectify biases, mistakes, or unexpected repercussions in real-world scenarios.

3 Biases in AI Algorithms in Healthcare

Artificial intelligence algorithms have the capacity to revolutionize healthcare delivery by facilitating more precise diagnoses, tailored treatment regimens, and enhanced patient outcomes. Nevertheless, like many other technologies, AI algorithms are vulnerable to biases that might influence their efficiency and impartiality in healthcare environments. These biases may originate from several sources, such as skewed data utilized for training algorithms, deficiencies in algorithm design, and the unconscious prejudices of human developers or users. Gaining a comprehensive understanding of these biases is essential to guarantee that AI-powered healthcare solutions are fair, precise, and reliable. This section delves into the many forms of biases present in AI algorithms used in healthcare and examines their consequences for patient care.

3.1 Data Bias

Data bias arises when the data utilized to train AI systems lack representativeness with respect to the audience they are intended to cater to. This might result in distorted or inadequate portrayals of specific demographic groups, leading to prejudiced forecasts or recommendations. Data bias may originate from several sources, such as [20]:

- *Underrepresentation*: Certain demographic groups may be underrepresented in healthcare datasets due to historical biases, lack of access to healthcare services, or other factors. For instance, if a dataset utilized to train an AI algorithm for the purpose of detecting skin disorders predominantly consists of photos of patients with lighter skin tones, the system may exhibit reduced accuracy when applied to patients with darker skin tones, resulting in discrepancies in diagnosis and treatment.
- *Sampling Bias*: Sampling bias arises when the data utilized to train AI systems are not randomly sampled or fail to accurately reflect the target population. For instance, if a dataset used to train an AI algorithm for predicting patient outcomes primarily includes data from academic medical centers or urban areas, the algorithm may not generalize well to patients from rural or underserved communities, leading to disparities in healthcare delivery.
- *Labelling Bias*: Labelling bias refers to the presence of biased or erroneous labels or annotations applied to data points within a dataset. For instance, if a dataset utilized to train an AI system for pneumonia diagnosis consists of photos labelled by radiologists with varying views of what defines pneumonia, the machine may acquire biased patterns and generate erroneous predictions.

Data bias in healthcare can have profound consequences for the delivery of medical services. Biased algorithms can result in inequalities in the diagnosis, treatment, and overall outcomes of patients [21]. To address data bias, it is necessary to ensure that training datasets are varied, representative, and unbiased. Additionally, approaches should be implemented throughout the data collecting and pre-processing stages to reduce biases.

3.2 Algorithmic Bias

Algorithmic bias pertains to inherent biases in the design or execution of AI systems. This can occur due to factors such as the choice of features used to train the algorithm, the complexity of the algorithmic model, or the optimization criteria used to train the algorithm. Algorithmic biases can result in discriminatory outcomes or reinforce existing inequalities in healthcare. Some common types of algorithmic bias include [22]:

- *Feature Selection Bias*: Feature selection bias occurs when certain features or variables used to train an AI algorithm are disproportionately weighted or prioritized,

leading to biased predictions or recommendations. For example, if an AI algorithm for predicting patient outcomes places too much emphasis on demographic factors such as age or gender, it may lead to disparities in treatment recommendations.
- *Model Bias*: Model bias occurs when the structure or complexity of the AI algorithm itself leads to biased predictions or recommendations. For example, if an AI algorithm for diagnosing diabetic retinopathy is trained on images from patients with severe forms of the disease, it may struggle to accurately diagnose milder cases, leading to disparities in diagnosis and treatment.
- *Optimization Bias*: Optimization bias occurs when the optimization criteria used to train an AI algorithm are biased or do not align with the desired outcomes. For example, if an AI algorithm for predicting patient readmission rates is optimized to minimize costs rather than improve patient outcomes, it may lead to biased recommendations that prioritize cost savings over patient care.

Algorithmic bias can perpetuate stereotypes, reinforce societal inequalities, and lead to discriminatory outcomes in healthcare AI applications [23]. Addressing algorithmic bias requires careful consideration of algorithm design, feature selection, and optimization criteria, as well as implementing techniques to mitigate biases during the training and evaluation stages.

3.3 Human Biases

Human biases can also influence AI algorithms, as they are often developed and trained by humans who may have unconscious biases or assumptions. These biases can appear in several forms, including the choice of training data, the construction of algorithmic models, or the analysis of algorithmic results. Human biases can perpetuate stereotypes, reinforce societal inequalities, and lead to discriminatory outcomes in healthcare AI applications. Some common types of human biases include [24]:

- *Confirmation Bias*: These biases can appear in several forms, including the choice of training data, the construction of algorithmic models, or the analysis of algorithmic results. For example, if developers believe that certain demographic groups are more likely to engage in unhealthy behaviors, they may inadvertently introduce biases into AI algorithms that reinforce these stereotypes.
- *Cultural Bias*: Cultural bias occurs when individuals unconsciously prioritize or value certain cultural norms or values over others. For example, if developers are primarily from one cultural background, they may inadvertently introduce biases into AI algorithms that favor the healthcare practices or treatment preferences of that culture over others.
- *Implicit Bias*: Implicit bias occurs when individuals hold unconscious attitudes or stereotypes about certain demographic groups that can influence their judgments or decisions. For example, if developers have implicit biases about the capabilities or intelligence of certain demographic groups, they may inadvertently introduce

biases into AI algorithms that affect how patients from those groups are diagnosed or treated.

Human biases have the potential to impact every step of the AI development process, starting from the collecting and pre-processing of data to the design and assessment of algorithms [25]. Addressing human biases requires raising awareness of unconscious biases among developers and stakeholders, implementing diversity and inclusion initiatives in AI development teams, and incorporating ethical considerations into all stages of AI development and deployment.

3.4 *Examples of Biases in Healthcare AI Applications*

Artificial intelligence has increasingly become a part of healthcare, promising to revolutionize diagnosis, treatment, and patient care. However, like any technology, AI systems are susceptible to biases that can have significant impacts on healthcare outcomes. Biases in healthcare AI applications can originate from several sources, such as biased data, poor algorithms, and human decision-making. Within this segment, we have examined many instances of biases in healthcare AI applications and their effects on patient care and results.

Gender Bias in Diagnostic Imaging

One prominent example of bias in healthcare AI applications is gender bias in diagnostic imaging. Studies have demonstrated that artificial intelligence (AI) systems utilized for the analysis of medical imagery, such as X-rays and mammograms, can display gender-based biases [26]. A study published in the journal JAMA Network Open found that AI algorithms used to interpret chest X-rays were more likely to misclassify images from female patients compared to male patients [27]. This gender bias can lead to inaccurate diagnoses and delays in treatment for female patients, potentially resulting in adverse health outcomes. The impact of gender bias in diagnostic imaging extends beyond individual patients to broader healthcare disparities. Women already face challenges in receiving timely and accurate diagnoses for certain conditions, such as heart disease and breast cancer [28]. Gender bias in AI algorithms exacerbates these disparities, further marginalizing women and underscoring the importance of addressing biases in healthcare AI applications.

Racial Bias in Clinical Decision Support Systems

Racial prejudice is a notable concern in healthcare AI applications, namely in clinical decision support systems (CDSS). Clinical Decision Support Systems are artificial intelligence systems specifically created to aid healthcare practitioners in formulating diagnostic and treatment choices by analyzing patient data. Studies have demonstrated that these systems can display racial biases, resulting in inequalities in the provision of healthcare [29]. For instance, research discovered that a commonly employed commercial AI system for forecasting healthcare requirements overestimated the requirements of Black patients in comparison to White patients [27]. This racial bias can result in inadequate treatment and management of health conditions among Black patients, contributing to disparities in healthcare outcomes. Racial bias in CDSS reflects broader systemic issues of racism and discrimination in healthcare. Addressing biases in AI applications requires not only technical solutions but also systemic changes to address underlying inequalities and promote health equity for all patients, regardless of race or ethnicity [30].

Socioeconomic Bias in Predictive Analytics

Socioeconomic bias is another prevalent issue in healthcare AI applications, particularly in predictive analytics models. Predictive analytics models use patient data to predict health outcomes and allocate resources efficiently. However, these models can exhibit biases based on socioeconomic status, leading to disparities in healthcare access and outcomes [31]. For example, a study found that predictive analytics models used to identify high-risk patients for care management programs were more likely to select patients from higher socioeconomic backgrounds [32]. This socioeconomic bias can result in disparities in access to care management programs and exacerbate existing healthcare inequalities. Addressing socioeconomic bias in predictive analytics requires a holistic approach that considers social determinants of health and incorporates data from diverse socioeconomic backgrounds. By accounting for socioeconomic factors in AI algorithms and ensuring equitable access to healthcare resources, stakeholders can mitigate biases and promote fairness in healthcare delivery.

Language Bias in Natural Language Processing

Language bias is a form of prejudice that may be found in healthcare AI applications, namely in natural language processing (NLP) systems. Natural Language Processing systems are employed for the purpose of examining and comprehending unorganized textual information, such as medical notes and patient records. Nevertheless, these systems may display biases influenced by linguistic and cultural elements, resulting in mistakes and discrepancies in healthcare results [33]. For instance, a research study discovered that natural language processing systems utilized for extracting information from electronic health records (EHRs) exhibited lower levels of accuracy when dealing with non-English-speaking patients in comparison to English-speaking patients [34]. The presence of language bias can lead to the misunderstanding of clinical data and the occurrence of mistakes in diagnosing and recommending treatments for patients who do not speak English. This emphasizes the need to remove biases in healthcare AI applications that span different languages and cultures.

Impact of Biases in Healthcare AI Applications

The consequences of biases in healthcare AI applications are extensive and can have significant repercussions for patient care and results. Biases can lead to inaccurate diagnoses, inappropriate treatment recommendations, and disparities in access to care, exacerbating existing healthcare inequalities and perpetuating systemic injustices [35]. Likewise, biases in AI applications undermine trust in healthcare systems and erode patient confidence in AI technologies, hindering their potential to improve healthcare delivery. The incorporation of AI algorithms in healthcare shows significant potential for better patient outcomes, optimizing treatment strategies, and improving the overall efficiency of healthcare provision. Nevertheless, the inherent biases found in AI systems pose substantial obstacles to establishing fair and impartial healthcare delivery. Prejudices can originate from several causes, such as the utilization of biased data for training algorithms, poor design of algorithms, and unconscious prejudices shown by developers or users [36]. It is crucial to confront these prejudices to guarantee the provision of healthcare that is just and impartial. During this conversation, we have examined the difficulties presented by biases in AI algorithms and how they affect the fair provision of healthcare.

AI algorithm biases might impact the precision of diagnosis and treatment suggestions, resulting in inequalities in healthcare provision [37]. According to research, an AI system used for identifying skin cancer had worse accuracy rates for individuals with darker skin tones compared to those with lighter skin tones [38]. This bias can result in delayed or incorrect diagnoses for patients from marginalized racial or ethnic groups, exacerbating health disparities and reducing access to timely and appropriate care. Likewise, prejudices in treatment recommendation algorithms might result in inequalities in the therapy choices shown to patients. An instance of this can be seen in a research article published in Health Affairs, which revealed that a machine learning algorithm employed to assist in treatment choices for patients with intricate medical conditions tended to suggest less aggressive treatments for Black patients in comparison to White patients, even after accounting for clinical factors [39]. This bias can lead to inequitable access to potentially life-saving medications and contribute to inequalities in health outcomes.

Impact on Access to Care

AI algorithms can have biases that impact the ability of disadvantaged and underserved groups to get care [40]. For instance, predictive analytics algorithms that are employed to identify patients who are likely to be readmitted to the hospital may exhibit biases that unfairly categorize individuals from low-income or minority populations as high-risk. This can result in heightened scrutiny and the possibility of being denied care. This can lead to obstacles in getting healthcare services for these groups and perpetuate the current inequalities in healthcare access and usage. Similarly, biases in AI algorithms used for scheduling appointments or allocating resources in healthcare settings can lead to inequitable distribution of services. A study found that an algorithm used to prioritize patients for organ transplantation tended to favor patients from higher socioeconomic status backgrounds, leading to disparities in access to transplantation for patients from disadvantaged communities [41]. This bias can further exacerbate existing disparities in access to specialized care and life-saving treatments.

Impact on Health Outcomes

Biases in AI algorithms can have significant implications for health outcomes, particularly for marginalized and underserved populations. For example, biases in AI algorithms used to predict patient outcomes or recommend treatment options may lead to suboptimal care and poorer health outcomes for patients from minority or disadvantaged backgrounds. This can result in higher rates of preventable complications, hospital readmissions, and mortality among these populations, further widening health disparities and perpetuating inequities in healthcare delivery [36]. Similarly, biases in AI algorithms used for population health management and resource allocation may lead to inadequate allocation of resources and interventions for communities with higher healthcare needs. For example, biases in predictive analytics algorithms used to identify communities at risk of disease outbreaks or public health emergencies may result in underestimation of risks in marginalized or underserved communities, delaying or limiting the deployment of resources and interventions to prevent or mitigate health crises [42].

Impact on Trust and Patient-Provider Relationships

Biases in AI algorithms can erode trust in healthcare systems and undermine patient-provider relationships. When patients perceive that AI algorithms are biased or discriminatory, they may be reluctant to trust the recommendations or decisions made by these algorithms, leading to decreased adherence to treatment plans and poorer health outcomes. Similarly, healthcare providers may lose trust in AI algorithms if they perceive that the algorithms are biased or unreliable, leading to hesitancy in relying on AI-driven decision support tools or recommendations [43]. Biases in AI algorithms can contribute to disparities in patient-provider interactions and communication. For example, biases in natural language processing algorithms used to analyze clinical notes or patient-provider communication may result in misinterpretation of patient symptoms or concerns, leading to suboptimal diagnosis and treatment. This can erode the quality of treatment offered to patients and contribute to inequities in the delivery of healthcare [44].

Addressing biases in AI algorithms is essential for achieving equitable healthcare delivery. This necessitates a comprehensive strategy that includes enhancing the accuracy and inclusiveness of data utilized for algorithm training, employing fairness-aware machine learning methods to address biases in algorithmic decision-making, fostering diversity and inclusivity in AI development teams, and improving transparency and accountability in the design and implementation of AI-driven healthcare systems. By addressing biases in AI algorithms, healthcare organizations can promote fair and equitable access to high-quality care for all patients, regardless of their demographic characteristics or social determinants of health [45].

4 Addressing Ethical Concerns and Biases in AI Implementation

As the incorporation of AI in healthcare progresses, it is crucial to acknowledge ethical issues and prejudices in order to guarantee that AI-powered healthcare systems uphold principles of impartiality, equitable treatment, and patient-focused care. A comprehensive strategy is needed to mitigate biases, enhance transparency and accountability, and safeguard patient privacy in AI implementation. This strategy should prioritize ethical principles, stringent regulations, and stakeholder involvement to promote trust and ensure the responsible use of AI in healthcare [46]. Proposal for a Comprehensive Strategy:

- Enhancing Data Quality and Representativeness: To tackle biases in AI algorithms, it is crucial to ensure that the training data used to create these algorithms is varied, representative, and devoid of biases. This involves collecting data from diverse populations, including underrepresented groups, and implementing measures to mitigate biases in data collection and pre-processing. Additionally, efforts should be made to address historical biases present in existing datasets to prevent the perpetuation of inequalities in AI-driven healthcare [47].
- Algorithmic Transparency and Explainability: Ensuring openness and comprehensibility in AI algorithms is crucial for cultivating trust and responsibility in AI-powered healthcare systems. Healthcare providers should prioritize the use of algorithms that are transparent and understandable to clinicians and patients. This may involve machine learning models, providing clear explanations of algorithmic outputs, and disclosing information about the underlying data and algorithms used [48].
- Fairness and Equal Treatment: Prioritizing fairness and equal treatment in AI-driven healthcare requires implementing measures to detect and mitigate biases in algorithms and decision-making processes [49]. This may involve conducting bias audits and assessments to identify potential sources of bias, implementing fairness-aware machine learning techniques to mitigate biases in algorithms, and regularly monitoring algorithm performance to ensure equitable treatment for all patients.
- Patient-Centric Care: Centering AI-driven healthcare systems on the needs and preferences of patients is essential for promoting patient-centric care. This entails engaging patients in the creation and advancement of AI-powered healthcare solutions, guaranteeing that patients can access comprehensible information regarding the utilization of AI algorithms in their treatment, and enabling patients to make well-informed choices about their healthcare treatment alternatives [50].
- Stringent Regulations and Standards: Implementing stringent regulations and standards for data management and algorithm evaluation is crucial for ensuring the responsible use of AI in healthcare. Regulatory bodies should establish guidelines and requirements for the collection, storage, and sharing of healthcare data,

as well as criteria for evaluating the performance, safety, and efficacy of AI algorithms. Compliance with these regulations should be mandatory for healthcare providers and AI developers to ensure patient safety and privacy [51].

- Ethical Oversight and Stakeholder Involvement: Establishing ethical oversight mechanisms and involving stakeholders in decision-making processes are essential for promoting accountability and transparency in AI-driven healthcare. This may involve establishing independent ethics committees or review boards to evaluate the ethical implications of AI-driven healthcare initiatives, engaging with patients, clinicians, policymakers, and other stakeholders to solicit feedback and input on AI implementation strategies, and ensuring that ethical considerations are integrated into all stages of AI development and deployment [52].

5 Recommendations for Mitigating Bias and Promoting Fairness in Healthcare AI

Given the growing use of AI technologies in the healthcare business, it is essential to prioritize efforts to reduce biases and ensure fairness in AI-driven systems. Addressing biases requires a multi-faceted approach that involves diversifying individuals involved in AI development, promoting diversity and inclusion in healthcare AI research and development, implementing strategies for evaluating and mitigating biases in AI algorithms, and ensuring on-going monitoring and adaptation to maintain fairness in healthcare AI systems.

One key strategy for mitigating biases in healthcare AI is to diversify the individuals involved in AI development. Studies have demonstrated that heterogeneous teams are more proficient at recognizing and resolving biases when compared to homogeneous teams. Incorporating individuals with various backgrounds, experiences, and viewpoints into AI development teams allows healthcare companies to mitigate the potential introduction of biases and guarantee that AI systems are built to be inclusive and equitable [53]. It is crucial to foster diversity and inclusivity in the research and development of healthcare AI on a larger scale, in addition to the efforts made to diversify AI development teams. This includes ensuring that research studies and datasets used to train AI algorithms are representative of the diverse populations they are meant to serve. Researchers should actively seek out data from underrepresented groups and strive to address gaps and biases in existing datasets. Furthermore, healthcare organizations should prioritize funding and support for research projects that focus on addressing health disparities and promoting health equity [54].

To effectively mitigate biases in healthcare AI algorithms, it is essential to implement strategies for evaluating and mitigating biases throughout the development lifecycle. This includes conducting bias assessments and audits to identify potential sources of bias in AI algorithms and implementing techniques to mitigate biases in algorithmic decision-making processes. Some strategies for evaluating and mitigating biases in AI algorithms include:

- *Bias Audits*: Conducting bias audits to identify potential sources of bias in AI algorithms and decision-making processes. Bias audits entail the analysis of training data, algorithmic models, and decision outputs to detect patterns of bias and evaluate the influence of biases on healthcare results [55].
- *Fairness-Aware Machine Learning*: Implementing fairness-aware machine learning techniques to mitigate biases in AI algorithms. Fairness-aware machine learning techniques aim to ensure that AI algorithms produce fair and equitable outcomes for all individuals, regardless of their demographic characteristics.
- *Bias Mitigation Strategies*: Implementing bias mitigation strategies such as data augmentation, bias correction techniques, and algorithmic adjustments to reduce the impact of biases in AI algorithms. These strategies aim to minimize the influence of biased data or decision-making processes on algorithmic outputs and promote fair and equitable outcomes [56].
- *Algorithmic Transparency and Explainability*: Enhancing algorithmic transparency and explainability to enable stakeholders to understand how AI algorithms work, what factors influence their decisions, and how biases are addressed. Transparent and explainable AI algorithms allow clinicians and patients to trust and validate algorithmic outputs and ensure that biases are appropriately addressed [57].

Finally, it is essential to emphasize the importance of on-going monitoring and adaptation to ensure fairness in healthcare AI systems. Healthcare organizations should implement processes for continuously monitoring AI algorithms and decision-making processes for biases and adverse effects. This includes monitoring algorithm performance, collecting feedback from clinicians and patients, and conducting regular audits and evaluations of AI systems [58]. Additionally, healthcare organizations should prioritize adaptability and flexibility in AI systems to allow for adjustments and refinements based on feedback and new information.

6 Conclusion

The incorporation of artificial intelligence into the healthcare sector has the capacity to profoundly transform the provision of patient care and enhance healthcare results. However, this transformation must be undertaken with careful consideration of key ethical considerations and biases inherent in AI applications. Safeguarding patient privacy, obtaining informed consent, ensuring transparency, and addressing biases are critical aspects that require attention. It is paramount for stakeholders in healthcare to prioritize fairness, transparency, and patient well-being in AI implementation to maximize the benefits of AI technology while ensuring equitable treatment for all patients. By actively and cooperatively tackling these problems, those involved may establish confidence, encourage responsibility, and cultivate the ethical utilization of AI in the healthcare sector, ultimately resulting in enhanced patient results and superior healthcare provision for everyone.

References

1. Ahmed, Z., Mohamed, K., Zeeshan, S., Dong, X.: Artificial intelligence with multi-functional machine learning platform development for better healthcare and precision medicine. Database 2020(baaa010) (2020). https://doi.org/10.1093/database/baaa010
2. Lee, D., Yoon, S.N.: Application of artificial intelligence-based technologies in the healthcare industry: opportunities and challenges. Int. J. Environ. Res. Public Health **18**(1), 271 (2021). https://doi.org/10.3390/ijerph18010271
3. Panayides, A.S., et al.: AI in medical imaging informatics: current challenges and future directions. IEEE J. Biomed. Health Inform. **24**(7), 1837–1857 (2020). https://doi.org/10.1109/JBHI.2020.2991043
4. Alowais, S.A., et al.: Revolutionizing healthcare: the role of artificial intelligence in clinical practice. BMC Med. Educ. **23**(689) (2023). https://doi.org/10.1186/s12909-023-04698-z
5. Paul, D., et al.: Artificial intelligence in drug discovery and development. Drug Discovery Today **26**(1), 80–93 (2021). https://doi.org/10.1016/j.drudis.2020.10.010
6. Yang, C.C.: Explainable artificial intelligence for predictive modeling in healthcare. J. Healthc. Inf. Res. **6**, 228–239 (2022). https://doi.org/10.1007/s41666-022-00114-1
7. Liefaard, M.C., et al.: The way of the future: personalizing treatment plans through technology. Am. Soc. Clin. Oncol. Educ. Book **41**, 12–23 (2021). https://doi.org/10.1200/EDBK_320593
8. Abràmoff, M.D., et al.: Considerations for addressing bias in artificial intelligence for health equity. npj Digital Med. 6(170) (2023). https://doi.org/10.1038/s41746-023-00913-9
9. Karimian, G., et al.: The ethical issues of the application of artificial intelligence in healthcare: a systematic scoping review. AI Ethics **2**, 539–551 (2022). https://doi.org/10.1007/s43681-021-00131-7
10. Zhang, H., et al.: Patient privacy and autonomy: a comparative analysis of cases of ethical dilemmas in China and the United States. BMC Med. Ethics **22**, 1–8 (2021). https://doi.org/10.1186/s12910-021-00579-6
11. Abouelmehdi, K., et al.: Big healthcare data: preserving security and privacy. J. Big Data 5(1) (2018). https://doi.org/10.1186/s40537-017-0110-7
12. Reddy, S., et al.: A governance model for the application of AI in health care. J. Am. Med. Inform. Assoc. **27**(3), 491–497 (2020). https://doi.org/10.1093/jamia/ocz192
13. Lorenzini, G., Shaw, D.M., Arbelaez Ossa, L., Elger, B.S.: Machine learning applications in healthcare and the role of informed consent: ethical and practical considerations. Clin. Ethics **18**(4), 451–456 (2023). https://doi.org/10.1177/14777509221094476
14. Zhang, Z., Genc, Y., Wang, D., et al.: Effect of AI explanations on human perceptions of patient-facing AI-powered healthcare systems. J. Med. Syst. **45**, 64 (2021). https://doi.org/10.1007/s10916-021-01743-6
15. Esmaeilzadeh, P.: Use of AI-based tools for healthcare purposes: a survey study from consumers' perspectives. BMC Med. Inform. Decis. Mak. **20**(1), 170 (2020). https://doi.org/10.1186/s12911-020-01191-1
16. de Fine Licht, K., de Fine Licht, J.: Artificial intelligence, transparency, and public decision-making. AI Soc. **35**, 917–926 (2020). https://doi.org/10.1007/s00146-020-00960-w
17. Ueda, D., Kakinuma, T., Fujita, S., et al.: Fairness of artificial intelligence in healthcare: review and recommendations. Jpn. J. Radiol. **42**, 3–15 (2024). https://doi.org/10.1007/s11604-023-01474-3
18. Bleher, H., Braun, M.: Diffused responsibility: attributions of responsibility in the use of AI-driven clinical decision support systems. AI Ethics **2**, 747–761 (2022). https://doi.org/10.1007/s43681-022-00135-x
19. Asan, O., Bayrak, A.E., Choudhury, A.: Artificial intelligence and human trust in healthcare: focus on clinicians. J. Med. Internet Res. **22**(6), e15154 (2020). https://doi.org/10.2196/15154
20. Celi, L. A., Cellini, J., Charpignon, M. L., Dee, E. C., Dernoncourt, F., Eber, R., Mitchell, W.G., Moukheiber, L., Schirmer, J., Situ, J., Paguio, J., Yao, S.: Sources of bias in artificial intelligence that perpetuate healthcare disparities—A global review. PLOS Digital Health **1**(3), e0000022 (2022). https://doi.org/10.1371/journal.pdig.0000022

21. Ntoutsi, E., Fafalios, P., Gadiraju, U., et al.: Bias in data-driven artificial intelligence systems—An introductory survey. WIREs Data Min. Knowl. Discovery **10**, e1356 (2020). https://doi.org/10.1002/widm.1356
22. Agarwal, R., Bjarnadottir, M., Rhue, L., Dugas, M., Crowley, K., Clark, J., Gao, G.: Addressing algorithmic bias and the perpetuation of health inequities: an AI bias aware framework. Health Policy Technol. **12**(1), 100702 (2023). https://doi.org/10.1016/j.hlpt.2022.100702
23. Panch, T., Mattie, H., Atun, R.: Artificial intelligence and algorithmic bias: implications for health systems. J. Glob. Health **9**(2), 010318 (2019). https://doi.org/10.7189/jogh.09.020318
24. Ferrara, E.: Fairness and bias in artificial intelligence: a brief survey of sources, impacts, and mitigation strategies. Science **6**(1), 3 (2023)
25. Howard, A., Borenstein, J.: The ugly truth about ourselves and our robot creations: the problem of bias and social inequity. Sci. Eng. Ethics **24**, 1521–1536 (2018). https://doi.org/10.1007/s11948-017-9975-2
26. Cirillo, D., Catuara-Solarz, S., Morey, C., et al.: Sex and gender differences and biases in artificial intelligence for biomedicine and healthcare. NPJ Digital Med. **3**, 81 (2020). https://doi.org/10.1038/s41746-020-0288-5
27. Jain, A., Brooks, J.R., Alford, C.C., Chang, C.S., Mueller, N.M., Umscheid, C.A., Bierman, A.S.: Awareness of racial and ethnic bias and potential solutions to address bias with use of health care algorithms. JAMA Health Forum **4**(6), e231197–e231197 (2023). https://doi.org/10.1001/jamahealthforum.2023.1197
28. Gorin, S.S., Heck, J.E., Cheng, B., Smith, S.J.: Delays in breast cancer diagnosis and treatment by racial/ethnic group. Arch. Intern. Med. **166**(20), 2244–2252 (2006). https://doi.org/10.1001/archinte.166.20.2244
29. Koski, E., Scheufele, E.L., Karunakaram, H., Foreman, M.A., Felix, W., Dankwa-Mullan, I.: Understanding disparities in healthcare: implications for health systems and AI applications. In: Healthcare Information Management Systems: Cases, Strategies, and Solutions, pp. 375–387. Springer, Cham (2022)
30. Briggs, E., Hollmén, J.: Mitigating discrimination in clinical machine learning decision support using algorithmic processing techniques. In: Appice, A., Tsoumakas, G., Manolopoulos, Y., Matwin, S. (eds.) Discovery Science. DS 2020. Lecture Notes in Computer Science, vol. 12323. Springer, Cham (2020). https://doi.org/10.1007/978-3-030-61527-7_2
31. Juhn, Y.J., Ryu, E., Wi, C.I., King, K.S., Malik, M., Romero-Brufau, S., Weng, C., Sohn, S., Sharp, R.R., Halamka, J.D.: Assessing socioeconomic bias in machine learning algorithms in health care: a case study of the HOUSES index. J. Am. Med. Inf. Assoc. **29**(7), 1142–1151. https://doi.org/10.1093/jamia/ocac052
32. Norori, N., Hu, Q., Aellen, F.M., Faraci, F.D., Tzovara, A.: Addressing bias in big data and AI for health care: a call for open science. Patterns **2**(10) (2021). https://doi.org/10.1016/j.patter.2021.100347
33. Li, I., Pan, J., Goldwasser, J., Verma, N., Wong, W.P., Nuzumlalı, M.Y., Rosand, B., Li, Y., Zhang, M., Chang, D., Taylor, R.A., Radev, D.: Neural natural language processing for unstructured data in electronic health records: a review. Comput. Sci. Rev. **46**, 100511 (2022). https://doi.org/10.1016/j.cosrev.2022.100511
34. Flores, G.: Language barriers and hospitalized children: are we overlooking the most important risk factor for adverse events? JAMA Pediatr. **174**(12), e203238 (2020). https://doi.org/10.1001/jamapediatrics.2020.3238
35. Zhang, J., Zhang, Z.: Ethics and governance of trustworthy medical artificial intelligence. BMC Med. Inform. Decis. Mak. **23**(1), 7 (2023). https://doi.org/10.1186/s12911-023-02103-9
36. Chin, M.H., Afsar-Manesh, N., Bierman, A.S., et al.: Guiding principles to address the impact of algorithm bias on racial and ethnic disparities in health and health care. JAMA Netw. Open **6**(12), e2345050 (2023). https://doi.org/10.1001/jamanetworkopen.2023.45050
37. Chen, R.J., Wang, J.J., Williamson, D.F.K., et al.: Algorithmic fairness in artificial intelligence for medicine and healthcare. Nature Biomed. Eng. **7**, 719–742 (2023). https://doi.org/10.1038/s41551-023-01056-8

38. Rezk, E., Eltorki, M., El-Dakhakhni, W.: Leveraging artificial intelligence to improve the diversity of dermatological skin color pathology: protocol for an algorithm development and validation study. JMIR Res. Protocols **11**(3), e34896 (2022). https://doi.org/10.2196/34896
39. Sun, M., Oliwa, T., Peek, M.E., Tung, E.L.: Negative patient descriptors: documenting racial bias in the electronic health record: study examines racial bias in the patient descriptors used in the electronic health record. Health Aff. **41**(2), 203–211 (2022)
40. Delgado, J., de Manuel, A., Parra, I., et al.: Bias in algorithms of AI systems developed for COVID-19: a scoping review. Bioethical Inquiry **19**, 407–419 (2022). https://doi.org/10.1007/s11673-022-10200-z
41. Venugopal, S., Stoner, E., Cadeiras, M., et al.: Understanding organ transplantation in the USA using geographical social networks. Soc. Netw. Anal. Min. **3**, 457–473 (2013). https://doi.org/10.1007/s13278-012-0089-1
42. Nazer, L.H., Zatarah, R., Waldrip, S., et al.: Bias in artificial intelligence algorithms and recommendations for mitigation. PLOS Digital Health **2**(6), e0000278 (2023). https://doi.org/10.1371/journal.pdig.0000278
43. Adler-Milstein, J., Aggarwal, N., Ahmed, M., et al.: Meeting the moment: addressing barriers and facilitating clinical adoption of artificial intelligence in medical diagnosis. NAM Perspect. **2022** (2022). https://doi.org/10.31478/202209c. https://doi.org/10.31478/202209c
44. Liu, L.G. The Transparent, Concurrent, and Collaborative Health Record: Methods to Improve Patients' Comprehension of Health Information. Columbia University (2021)
45. Thomasian, N.M., Eickhoff, C., Adashi, E.Y.: Advancing health equity with artificial intelligence. J. Public Health Policy **42**, 602–611 (2021). https://doi.org/10.1057/s41271-021-00319-5
46. Monlezun, D.J. The Thinking Healthcare System: Artificial Intelligence and Human Equity. Elsevier (2023)
47. Osoba, O.A., Welser IV, W., Welser, W.: An Intelligence in Our Image: The Risks of Bias and Errors in Artificial Intelligence. Rand Corporation (2017)
48. Shin, D.: User perceptions of algorithmic decisions in the personalized AI system: perceptual evaluation of fairness, accountability, transparency, and explainability. J. Broadcast. Electron. Media **64**(4), 541–565 (2020). https://doi.org/10.1080/08838151.2020.1843357
49. Drukker, K., Chen, W., Gichoya, J.W., Gruszauskas, N. P., Kalpathy-Cramer, J., Koyejo, S., Myers, K., Sá, R.C., Sahiner, B., Whitney, H., Zhang, Z., Giger, M.L. J. Med. Imaging **10**(6), 061104 (2023). https://doi.org/10.1117/1.JMI.10.6.061104
50. Aminabee, S.: The future of healthcare and patient-centric care: digital innovations, trends, and predictions. In: Garcia, M., de Almeida R. (eds.) Emerging Technologies for Health Literacy and Medical Practice, pp. 240–262. IGI Global (2024). https://doi.org/10.4018/979-8-3693-1214-8.ch012
51. Reddy, S., Rogers, W., Makinen, V.P., Coiera, E., Brown, P., Wenzel, M., Ansari, S., Mathur, P., Casey, A., Kelly, B.: Evaluation framework to guide implementation of AI systems into healthcare settings. BMJ Health Care Inf. **28**(1), e100444 (2021). https://doi.org/10.1136/bmjhci-2021-100444
52. Elendu, C.M., Amaechi, D.C., Elendu, T.C., Jingwa, K.A., Okoye, O.K., John Okah, M., Ladele, J.A., Farah, A.H., Alimi, H.A.: Ethical implications of AI and robotics in healthcare: a review. Medicine **102**(50), e36671 (2023). https://doi.org/10.1097/MD.0000000000036671
53. Gichoya, J.W., Thomas, K., Celi, L.A., Safdar, N., Banerjee, I., Banja, J.D., Seyyed-Kalantari, L., Trivedi, H., Purkayastha, S.: AI pitfalls and what not to do: mitigating bias in AI. Br. J. Radiol. **96**(1150), 20230023 (2023). https://doi.org/10.1259/bjr.20230023
54. Mohamed, S., Png, M.T., Isaac, W.: Decolonial AI: decolonial theory as sociotechnical foresight in artificial intelligence. Philos. Technol. **33**, 659–684 (2020). https://doi.org/10.1007/s13347-020-00405-8
55. Daneshjou, R., Smith, M.P., Sun, M.D., Rotemberg, V., Zou, J.: Lack of transparency and potential bias in artificial intelligence data sets and algorithms: a scoping review. JAMA Dermatol. **157**(11), 1362–1369 (2021). https://doi.org/10.1001/jamadermatol.2021.3129

56. Balayn, A., Lofi, C., Houben, G.J.: Managing bias and unfairness in data for decision support: a survey of machine learning and data engineering approaches to identify and mitigate bias and unfairness within data management and analytics systems. VLDB J. **30**, 739–768 (2021). https://doi.org/10.1007/s00778-021-00671-8
57. Felzmann, H., Fosch-Villaronga, E., Lutz, C., et al.: Towards transparency by design for artificial intelligence. Sci. Eng. Ethics **26**, 3333–3361 (2020). https://doi.org/10.1007/s11948-020-00276-4
58. Lysaght, T., Lim, H.Y., Xafis, V., et al.: AI-assisted decision-making in healthcare. Asian Bioethics Rev. **11**, 299–314 (2019). https://doi.org/10.1007/s41649-019-00096-0

Azamat Ali is an Assistant Professor in the School of Liberal and Creative Arts (Journalism and Mass Communication) at Lovely Professional University, Punjab, India. He holds an M.Phil. and Ph.D. from the Central University of Gujarat, Gandhinagar. Additionally, he earned a B.Sc. in Physics (Hons.) from Aligarh Muslim University, an M.Sc. in Science Communication from Devi Ahilya University, Indore, and a Postgraduate Diploma in Broadcast Technology from Jamia Millia Islamia, New Delhi. Dr. Ali's research interests encompass emerging technologies, development communication, and health and science communication. He has made notable contributions to these fields, with over 20 research papers published in reputed journals and several book chapters in Scopus-indexed publications. He has also presented at numerous national and international conferences and seminars, consistently advancing the dialogue on science, technology, and innovation. Dr. Ali's work is focused on bridging the gap between technological advancements and societal benefits, emphasizing the transformative role of emerging technologies.

A Comprehensive Review of Encryption and Protection Techniques for Healthcare Data

Venkata Nedunoori

Abstract In today's digital age, there are a vast number of studies and extensive research being conducted to advance the healthcare standards of individuals in all areas of healthcare domains. The healthcare industry is constantly evolving and has witnessed a remarkable technological advancement in transforming patient care, diagnosis, treatments methods and pharmaceutical domains. Some of the key areas where technology has made a significant impact is Telemedicine, Electronic Health Records, Wearable Technology, AI and ML, Robotics and 3D printing etc. With the advent of the emerging technological advancements in the healthcare industry, the need to protect healthcare data has become a greater concern. The key reasons are increased data volume, complexity, interoperability and data sharing. Apart from these, the need of regulatory compliance and maintain trust and privacy of the healthcare data weights in robust data security and protection mechanisms to be implemented to safeguard the healthcare information. The goal of this chapter is to provide a comprehensive review of data encryption and protection techniques for both data at rest and in transit. It explores the application-level encryption strategies that are critical in maintaining confidentiality and integrity in handling healthcare data. Along with topics related practical challenges such as key management and maintain a balance between data utility and privacy, the chapter also discusses emerging technologies and future trends in the realm of data security and protection in healthcare such as homomorphic encryption, blockchain, quantum cryptography and the integration of AI in automated data protection processes.

Keywords Healthcare data security · Data protection techniques · Telemedicine · Electronic health records · Wearable technology · Artificial intelligence (AI) in healthcare · Machine learning (ML) · Data encryption · Homomorphic encryption · Quantum cryptography

V. Nedunoori (✉)
Dentsu International, Frisco, TX, USA
e-mail: vnedunoori@gmail.com

N. R. Vajjhala et al. (eds.), *Artificial Intelligence in Healthcare Information Systems—Security and Privacy Challenges*, Information Systems Engineering and Management 34, https://doi.org/10.1007/978-3-031-84404-1_8

1 Introduction

In the rapidly evolving landscape of healthcare, technological advancements have significantly transformed patient care, diagnosis, treatment methods, and the broader medical field. From telemedicine to AI-driven diagnostics and personalized treatment plans, the integration of technology has opened unprecedented possibilities in healthcare. However, these advancements have also introduced significant vulnerabilities, particularly concerning data security and protection. The sensitive nature of healthcare data, which includes personal health information (PHI), financial details, and other confidential records, makes it a prime target for cyber threats. Data breaches in healthcare can lead to severe consequences, including compromised patient privacy, legal liabilities, regulatory penalties, and financial losses [1–3]. Beyond these immediate risks, data breaches can undermine the integrity of healthcare data, rendering it unreliable for medical research and clinical decision-making [4, 5]. As healthcare organizations increasingly rely on digital systems and interconnected medical devices, ensuring robust data security has become a critical priority.

At the heart of data security lies encryption, a fundamental technique that encodes information to make it inaccessible to unauthorized users. Encryption transforms readable data, or plaintext, into an unreadable format, known as ciphertext, which can only be deciphered with the appropriate key [6]. In the context of healthcare, encryption is essential not only for protecting data from unauthorized access but also for maintaining the confidentiality and integrity of information as it moves across networks and between systems. This chapter examines the essential role of encryption in safeguarding healthcare data. It explores the types of encryptions—symmetric and asymmetric—and their relevance to securing data both at rest and in transit. The chapter also examines advanced encryption strategies, such as Field-Level Encryption (FLE) and End-to-End Encryption (E2EE), which provide granular control over data security and ensure secure communications. As healthcare data becomes increasingly complex and widely utilized, the need for sophisticated encryption techniques grows. This chapter also addresses emerging technologies like Homomorphic Encryption and Blockchain, which are redefining data security by enabling secure data processing and ensuring immutable data integrity. Additionally, the chapter discusses practical challenges associated with implementing robust encryption strategies, including key management, performance overhead, and the ethical implications of data access and privacy.

Regulatory compliance is another critical aspect of data protection in healthcare. Legal frameworks such as HIPAA in the United States and GDPR in Europe not only mandate specific security measures but also shape the overall approach to data protection [2, 3]. Compliance with these regulations is essential for healthcare organizations to protect patient data and avoid legal repercussions. Finally, this chapter explores future trends and emerging technologies, such as Quantum Cryptography and the integration of Artificial Intelligence in automating data protection processes. These advancements are poised to significantly influence the future of healthcare data security, offering new tools and strategies to address evolving threats and challenges. This chapter uniquely contributes by not only reviewing established encryption methods but also by exploring innovative technologies like Homomorphic Encryption, Blockchain, and Quantum Cryptography. These emerging technologies are particularly relevant as they offer advanced solutions to the complex security and privacy challenges posed by AI-driven healthcare systems. Moreover, the chapter provides practical insights into managing key challenges, such as ensuring robust data security without compromising the accessibility and utility of healthcare data, making it an essential resource for both researchers and practitioners navigating the evolving landscape of healthcare information systems.

2 Core Principles and Types of Data Encryption in Healthcare

Data encryption is an effective technique to protect data security [6]. It serves as the foundation of data integrity and confidentiality by converting the human readable text into an unreadable random format that is not understandable known as ciphertext [7]. In the context of healthcare data, as the organizations handle huge amounts of personal and sensitive data that include health records and financial information. This makes the need for implementing and following robust encryption practices essential. In the further sections we will explore the types of encryption and common algorithms associated.

Figure 1 represents a plaintext and respective Ciphertext using an encryption key. The encryption key is a randomly generated key (using fernet) that will be used for both encryption and decryption in the case of symmetric encryption [8].

2.1 Principles for Robust Encryption Strategy

The key principles of encryption are:

Confidentiality Data confidentiality ensures that the sensitive data is not accessible to the unauthorized users and ensures that the data received by the receiver is consistent with the data sent by the sender [6]. Moreover, encryption protects data privacy

Plaintext: "Sensitive patient data: Patient ID 123456, Blood Type O+"

Ciphertext:
"gAAAAABmdzwBHV_PUF_v11_VBg7ALvGFbXvi9319jwaJ3iuHsm_812kWqLTlDW-pu93JgSEf0rZAOVZlgydkBXS4d5wY_rrD4ORTJLQBAM3cZt-VR-3qYwHeYajk-K0ijFqxzp-cLoXKwrD5hRK0laqbT3AFLjc8Nw=="

Fig. 1 Sample Plaintext and respective Ciphertext generated using symmetric encryption

by making sure that even if the data is intercepted, it cannot be decrypted without the necessary key.

Integrity Data integrity ensures that data is unaltered in unauthorized ways. Data integrity ensures the reliability of data [6]. Data encryption combined with data integrity mechanisms such as digital signatures and hash functions ensure the integrity of data.

Authentication Authentication is one of the aspects that ensures that the data originates from a trusted and expected source. This aids in validating the parties involved in the communication with respect to data sharing. Encryption facilitates authentication using digital certificates or certificate-based authentication.

Non-repudiation Non-repudiation prevents an entity from denying the authenticity of their signature on a message. Repudiating interactions (mainly during transmission of data or on their storage) is often counteracted by preventing authorized access in the first place. These techniques are therefore often issued to address access control requirements and are classified as such. Amongst others, they include the exchange of public keys (PKI), certificates or (proxy) signatures [9].

Key Management Key management is crucial to the security of the encryption system. The key management basically deals with the secure creation, distribution, storage as well as destruction of keys. Periodic key rotation and secure key management practices are essential for a robust encryption system [10].

Access Control Proper access control measures are essential to determine the systems/users that are authorized to encrypt or decrypt the data [11].

2.2 Comparative Analysis of Symmetric and Asymmetric Encryption

Encryption can be classified into two types at the high level: symmetric and asymmetric. The usage of type of encryption depends upon the use case within the context of the healthcare system.

Symmetric Encryption In this type of encryption, the same key that is used to encrypt the data is used to decrypt the data [12]. This makes the encryption faster and efficient. This type of encryption technique is ideal for encrypting larger volumes of data such as entire databases. However, the use of AES algorithm (Advanced Encryption Standard) is generally adopted in healthcare systems as it is considered a robust encryption algorithm (Fig. 2).

Fig. 2 Symmetric encryption

Fig. 3 Asymmetric encryption

Asymmetric Encryption In this type of encryption, two different keys are utilized for encryption and decryption of the data- Public key and Private key. Public is usually available with all the trust sources to encrypt the data, whereas the private key is kept as secret and only available to system that's intended to decrypt the data. RSA (Rivest-Shamir-Adleman) algorithm is generally considered for asymmetric type of encryption [12]. In healthcare context, usually this type of encryption is implemented in data in transit scenarios where the data is usually transferred between multiple systems such as healthcare providers and insurance companies (Fig. 3).

2.3 Evaluating Key Encryption Algorithms for Healthcare Data Security

The adaptation and implementation of proper algorithms is critical for robust data security. This section discusses key algorithms, and their suitability of healthcare data based on the specific use case.

Advanced Encryption Standard (AES) This is one of the robust symmetric encryption algorithms that can be mainly utilized for encrypting the data at rest. The speed and robustness of the algorithm makes it ideal for encrypting large databases [13].

Rivest-Shamir-Adleman (RSA) RSA is one of the earliest asymmetric encryption algorithms and that's still widely adopted for secure data transmission. In simple terms, it is used to encrypt the keys that are then used to encrypt/decrypt large data sets with symmetric methods like AES [14].

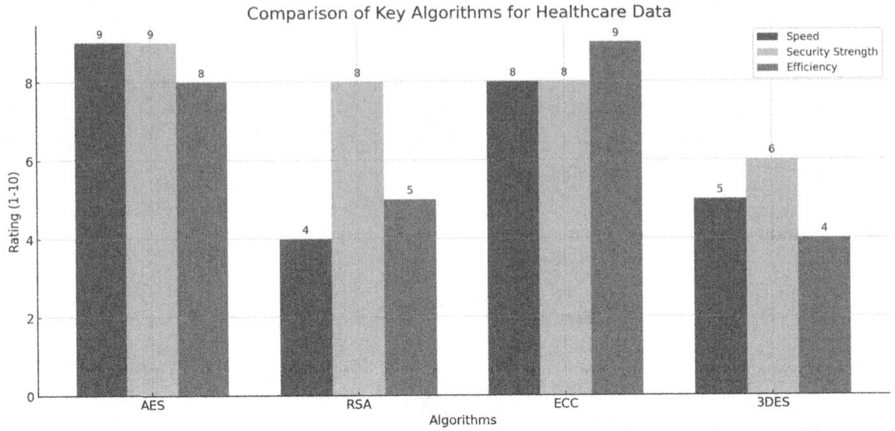

Fig. 4 Comparison of key algorithms

Elliptic Curve Cryptography (ECC) This encryption technique is similar levels of security to RSA but with smaller key sizes. The same level of security with smaller key sizes makes ECC utilize reduced processing power and storage compared to RSA [15]. This efficiency makes it ideal for usage in mobile health applications and telehealth services. As these both the use cases require the data to be transmitted fast and secure.

Triple Data Encryption Standard (3DES). Its relevant in the context to mention that 3DES has been used in the older healthcare systems as a symmetric encryption. However, this is not as strong as AES and the technique to encrypt the data three times makes it secure but slower than AES [16] (Fig. 4).

3 Strategies for Securing Data at Rest in Healthcare Systems

In the healthcare industry protecting sensitive data is a top priority. This is due to the confidential nature of the data as the healthcare data primarily consists of large volumes of PII (Personal Identifiable Information) and PHI (Personal Health Information). Moreover, in the digital age due to the wide adoption of cloud-based storage solutions, ensuring data security becomes even more complex due to the distributed nature of cloud storage and computing [17]. Apart from the technological perspective it's also critical to prevent unauthorized access and comply with regulations such as HIPAA and GDPR.

3.1 Effective Techniques for Protecting Data at Rest

Full Disk Encryption (FDE) The Full Disk Encryption (FDE) is critical in protecting all the data stored on a disk. This primarily to ensure that all the data partitions and the files are encrypted so that the data is only accessible with correct encryption keys. The FDE technique is particularly useful for physical servers in data centers or on client devices that access the cloud. Encryption of data on disk drives is implemented at the file system encryption and full disk encryption (bitstream) levels [18]. Full disk encryption secures data on disk drives with a single symmetric key [19]. Full disk encryption protects data in all areas of the disk drive, including areas outside the file system. Such data are the hidden files, swap files, file metadata, temporary files and caches, registry files, and boot sector data [20–23]. The FDE is important in the healthcare setup as it protects data from theft or loss. As devices such as laptops, mobile devices and external hard drives are commonly used in healthcare settings, FDE on these devices ensures that the data contained on them remains protected and inaccessible to unauthorized users. In the events of data breach as well, FDE significantly reduces the risk of sensitive information to be exposed as the data is encrypted and cannot be decrypted without necessary decryption keys. Also, FDE helps the healthcare organizations to comply with local regulations such as HIPAA and GDPR.

Database Encryption Database encryption involves encrypting data within a database at the column or row level. Sometimes also involves encrypting the entire database file on the disk. This method is essential for protecting sensitive patient information stored in centralized databases that healthcare providers access and manage [24]. Apart from protecting sensitive data and compliance with regulations Database encryption also helps in maintaining data integrity and confidentiality [25].

Database encryption can be implemented using three techniques.

Transparent Data Encryption (TDE) TDE is a method where encryption and decryption are performed automatically and transparently at the database or storage level. This way is the encryption is implemented without requiring changes to the application. This is useful in healthcare for securing entire databases without affecting the end-user experience [26].

Column-Level Encryption This is helpful in scenarios where the databases contain both sensitive and non-sensitive information. This way encryption can be applied only on columns that store sensitive information. It is useful in the cases where there is a need to improve the performance when working against the non-sensitive data [25].

File-Level Encryption In the scenarios where it's required to encrypt the entire database file system, the File-level encryption is helpful. This differs from full-disk encryption as file level encryption provides more granular control by only allowing specific files to be encrypted rather than all data on the drive. In healthcare context this type of encryption is usually implemented to encrypt electronic health records and medical images.

3.2 Best Practices for Implementing Data Protection in Rest

Robust Key Management It's crucial to use a secure key management process. Usually this involves a centralized key management server that handles key storage, backup, and lifecycle management. Systems are prone to permanent data loss in case of losing the keys and at the same way compromised keys lead to data breaches [27].

Security Audits Regular security audits play a key role in ensuring compliance with policies and regulations. Audits also help in identifying vulnerabilities and enforcing security policies.

Access Controls Strict access controls ensure that only authorized systems or users can access sensitive data. Database activity monitoring can also be combined to further strengthen the access controls. Role based access controls (RBAC) and least privilege is specifically important in the sensitive environments.

Performance Considerations Effective encryption of data at rest can promote data security but it's also essential to consider performance implications that can arise due to over engineering of encryption [28].

4 Securing Healthcare Data in Transit

Healthcare systems interacting with each other often require sensitive data to be transferred over networks. The security of the data while it is being transferred over the networks plays an essential role in maintaining robust data security. Data security aspects such as protecting data from unauthorized access, alteration and interception while its being transferred are covered under data protection in transit [29]. Data protection in transit only conveys the sensitive data transfer between the systems but also emails, file transfers and API communications between healthcare applications [30]. In the real world the complexity of data protection in transit arises where data usually traverses through public and private networks. The key challenges in data protection in transit are dealing with highly sensitive information and complex network environments along with regulatory requirements.

4.1 Techniques for Data Protection in Transit

Application Layer Encryption Transport layer security (TLS) is the most widely used protocol for securing the data as it travels across the networks. However, application encryption provides an additional layer of security when compared to the standard TLS. In application layer security the data is encrypted at the application level before it is sent from the application [31]. This way the data travels encrypted all through the entire transit until it reaches the destination system.

VPN and IPsec Virtual private networks (VPNs) provide a secure and encrypted for the data to travel across the internet. This way the confidentiality and integrity of the data is protected while the data is in transit. On the other hand, Internet Protocol Security (IPsec) is responsible for encrypting each IP packet of the data that's in transit throughout the session. The IPsec is widely used in VPNs for securing communication between the sites in healthcare networks [32].

Secure Health Protocols Another key aspect to discuss in securing the data in transit for healthcare industry specifically is Secure health protocols. These protocols help in aligning with healthcare industry specific security needs. Protocols such as HL7 (Health Level 7) is a set of international standards for the exchange, integration and working with electronic health information. Digital imaging and communication in medicine (DICOM) is another standard for handling information in medical imaging [33]. This standard includes a framework for communicating medical images and related data securely between systems [34]. Fast Healthcare Interoperability Resources (FHIR) is another key standard the describes the data formats and elements along with API for the exchange of health records [35].

4.2 Best Practices for Implementing Data Protection in Transit

Comprehensive Data Mapping and Risk Assessments It is essential to determine all the elements through which the data flows and touches across the network. A through data mapping and risk assessment can help largely in identifying potential security vulnerabilities.

Implementing Zero Trust Security Model The Zero Trust model works on the principle that no entity should be automatically trusted. The entity can either be internal or external. Micro segmentation in the Zero Trust model guides breaking the security parameters into smaller segments. Through this approach in case a breach occurs the impact can be limited to a small segment of the network [36].

Data Loss Prevention (DLP) Strategies The use of DLP tools provide real time monitoring of data movements across the network [37]. This practice makes it possible for immediate detection of any policy violations or unauthorized transmission of data. When sensitive information is detected to transmit out of the company network, DLPs can be configured to trigger an alert for the action to be taken. In this way DLPs improve the management of data with minimum risk and prevent data breaches [37].

Regular Security Audits and Penetration Testing It's a best practice to conduct scheduled and unscheduled security audits regularly to access the validity of security policies. These audits combined with review of access logs, user permissions and effectiveness of security controls aid in ensuring robust data protection in transit.

5 Application-Level Encryption in Healthcare

Application-level encryption is a complex approach in securing sensitive data. In the context of healthcare data this approach can play a crucial role where the confidentiality and privacy of the data are maintained. In application-level encryption, the data is encrypted in the application itself before its stored [38]. The data remains encrypted until retrieved by the application. This approach provides a robust security layer irrespective of the underlying storage or transmission mechanisms. Hyper Text Transfer Protocol Secure (HTTPS) is one of the common examples of application-level encryption which encrypts the data that's transferred between web server and web browser. There are two types of application-level encryption – Field-level encryption and End-to-End encryption.

5.1 *Field-Level Encryption (FLE)*

Field-level encryption offers selective encryption targeting specific data fields for encryption such as patient IDs, medical records or financial details. This approach helps in improving the overall system performance by encrypting only the most sensitive data. The FLE is implemented at the application code level and the application logic is updated to encrypt and decrypt the data as it is read from or written to the database. Field-level encryption have critical benefits in handling sensitive healthcare in this approach each field can be encrypted using unique key which makes managing these encryption keys critical. It is important to note that, even though FLE offers improved security strategy, it adds complexity in application design due to increased logic related to key management. However, benefits associated are:

Security at Granular Level By adopting this approach and encrypting the data at the field level, the solution offers an ability to implement security controls at the granular level. This approach also helps in designing and implementing access controls that are specific based on the sensitivity of data.

Compliance Risk By targeting the encryption of data at the field level, this approach offers healthcare organizations to meet compliance requirements specific to type of sensitive data such as PCI DSS for payment information or HIPAA for health information (Fig. 5).

5.2 *End-to-End Encryption (E2EE)*

In End-to-End encryption the data is encrypted on the origination system with the ability to be decrypted only the recipient. The intermediate nodes only transfer the

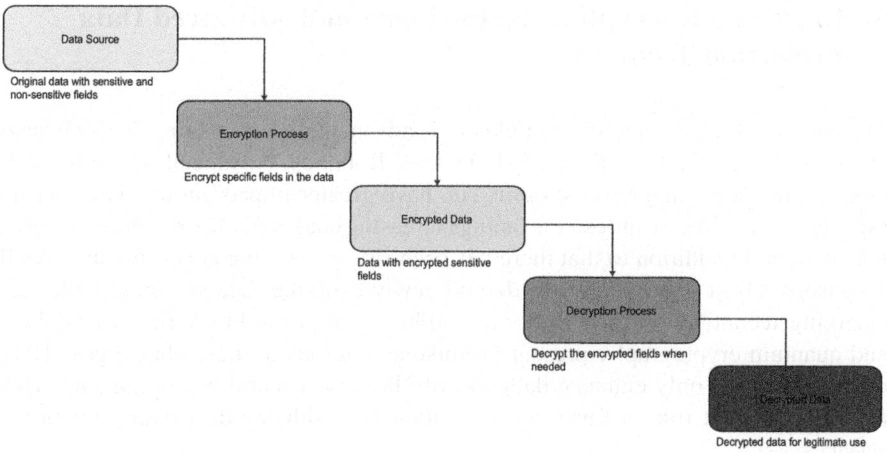

Fig. 5 Field level encryption process

data without the ability to decrypt [39]. This brings an advantage that even if the intermediate servers or nodes are compromised, the data remains protected. There by ensuring complete data privacy throughout the data transfer process and prevent any possible man-in-the-middle attacks [40]. This approach is important for handling sensitive data effectively where the criticality lies on the privacy and the security of patient communications. Implementing E2EE requires careful consideration of design to make sure the existing healthcare systems are compatible with this approach (Fig. 6).

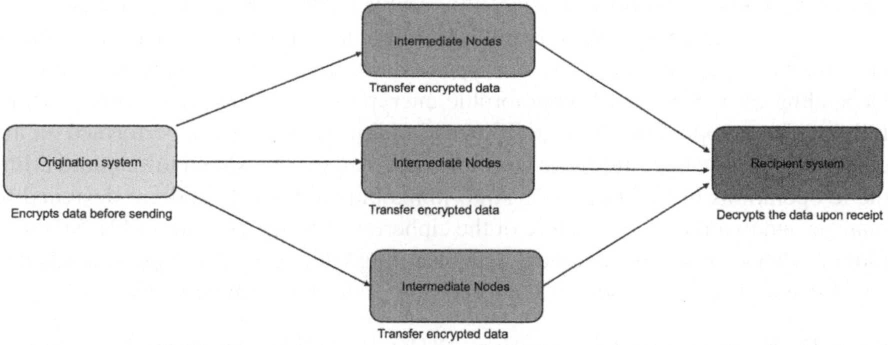

Fig. 6 End-to-end encryption process

6 Emerging Encryption Technologies and Advanced Data Protection Techniques

The pace at which the healthcare industry is advancing is astonishing. The advancements in terms of telemedicine, IoT devices, Robotics, Advanced Scanning techniques, innovative diagnosis systems etc. have greater impact on the data security aspects. The healthcare industry is facing increasing challenges in protecting sensitive patient data. In addition to that there has been rise in evolving cyber threats as well. To ensure robust data security to defend newly evolving data security challenges, emerging technologies such as homomorphic encryption, blockchain technology, and quantum cryptography present promising solutions to these challenges. These technologies not only enhance data security but also streamline compliance. They play a significant role in fostering innovation in healthcare data management and services.

6.1 Homomorphic Encryption

Homomorphic encryption is a major advancement in data security that's particularly important for industries like healthcare. Homomorphic encryption is a valuable approach where the confidentiality of patient information is crucial, yet there is need. For the data to be analyzed. This form of encryption offers the ability to perform analysis of the encrypted data and generate the results that once decrypted match the analysis that may have been performed on the original data [41]. This capability ensures sensitive data can remain securely encrypted throughout its processing, thus maintaining privacy while still being fully usable for the purpose of analysis.

Homomorphic encryption utilizes algorithms to perform arithmetic operations on ciphertexts, maintaining the encrypted state of data throughout the process. Depending on the type of homomorphic encryption used (partially, somewhat, or fully), different operations like addition and multiplication can be performed on the ciphertexts. These operations are carefully designed to correspond to standard arithmetic operations on the plaintext. Algorithms play a key role and are designed to manage and maintain the structure of the ciphertext during operations [42]. Methodologies such as Noise management, Ring learning with errors and bootstrapping can be adopted based on the type of homomorphic encryption implemented.

Partially Homomorphic Encryption (PHE) It provides the ability to perform a specific type of arithmetic operation, either addition or multiplication, on encrypted data without decrypting it. The key thing to note is it doesn't support both the addition and multiplication operations at the same time [42]. PHE stands out in terms of its simplicity and efficiency it brings to encryption. PHE is effective to be used in scenarios where analysis needs to aggregate encrypted data. Even though its simpler encryption mechanism to others, PHE has limitations because of single operation type. PHE is not suitable for more complex data processing tasks that require multiple types of computations.

Somewhat Homomorphic Encryption (SHE) SHE supports a certain limited number of both addition and multiplication operations on encrypted data. However, the system must at some point of time decrypt the results after a certain number of operations [42]. This is due to the growth in noise within the ciphertext that can result in decryption errors if not managed effectively. SHE is suitable for applications where a limited number of operations on encrypted data is enough to derive the results. Some practical use cases include performing a series of operations on encrypted data stored in the cloud, maintaining data privacy.

Fully Homomorphic Encryption (FHE) The most flexible type, allowing an unlimited number of operations on ciphertexts without compromising encryption. FHE is a type of data encryption that allows complex calculations to be performed on encrypted data without the need to decrypt it first [42]. This means that sensitive data can be processed in an encrypted form, ensuring it remains secure throughout its use, even from those doing the calculations [43]. The biggest benefit of FHE is privacy. Like SHE, FHE is also the most suitable for cloud computing and other services where sensitive information must be processed but may not fully trust the security of the system handling the data. For example, hospitals could use FHE to let researchers compute on patient data for studies without ever actually accessing patient records directly. This way, the data stays confidential and secure (Fig. 7).

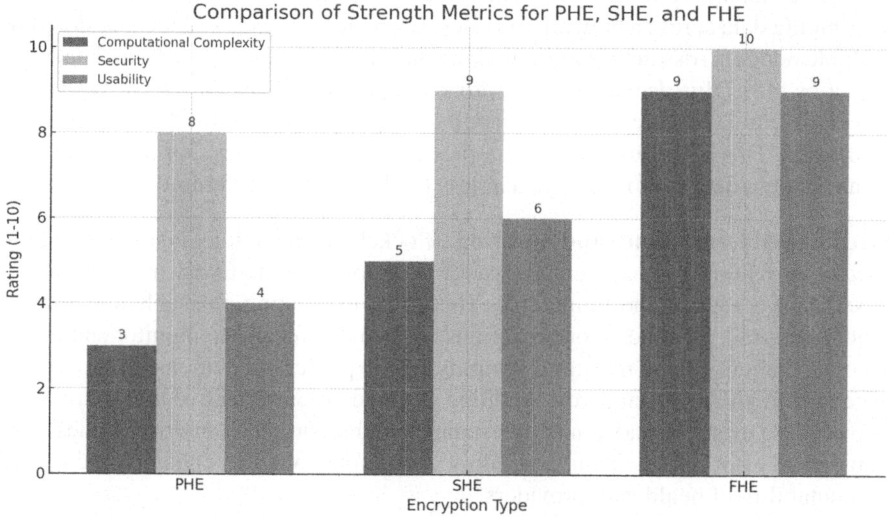

Fig. 7 Comparison of strength metrics for PHE, SHE and FHE

6.2 *Blockchain Technology for Enhancing Data Integrity and Security in Healthcare*

Blockchain technology is basically a digital ledger system that is used to record transactions across multiple computers. By this approach the records cannot be altered with effect from a date in the past [44]. This technology primarily applies to digital currencies like Bitcoin but has potential uses extending far beyond cryptocurrencies. Blockchain technology is increasingly being recognized and adopted for its ability to transform data management practices, particularly in sensitive fields such as healthcare. By using a decentralized and immutable ledger system, blockchain offers significant enhancements to the integrity and security of medical records and other critical data. Keyways in which Blockchain can bring significant impact to healthcare data management:

Immutable Patient Records One of the essential features of blockchain technology is its ability to create permanent and unchangeable records [45]. In healthcare, this means that every patient interaction with the healthcare system from doctor visits to treatment plans and medication schedules can be securely recorded on a blockchain. These records are timestamped and linked together in a chain that is cryptographically secured, ensuring that once a piece of data is added, it cannot be altered or deleted. This immutability is crucial for maintaining the integrity of medical history.

Decentralization of Data Storage Current data storage solutions usually centralize data in a single location. This can lead to a single point of failure and prone to cyberattacks or systemic failures. Blockchain technology addresses this vulnerability by distributing data across a network of nodes. Each piece of data, or block, is stored on multiple nodes across the network, making it highly resistant to attacks and data loss scenarios [45]. This decentralized approach greatly enhances data security. Even if one or several nodes are compromised, the integrity of the entire database remains unaffected. This is because identical copies exist on other nodes. For healthcare, this means patient data is more secure and less likely to be compromised.

Streamlined Compliance and Auditing Blockchain technology facilitates compliance by providing a transparent and easily auditable system. Every transaction on a blockchain is logged and immutable. This helps in creating a reliable and tamperproof audit trail. Healthcare organizations can use this feature to monitor and record access to patient data in real-time, simplifying the process of demonstrating compliance with regulatory standards. Auditing becomes more effective and more accurate because every transaction is timestamped and recorded on multiple nodes. This transparency not only helps in regulatory compliance but also enhances the overall accountability of healthcare providers.

6.3 Quantum Cryptography

Quantum cryptography is a significant advancement forward in the evolution of data security technologies. This is particularly with its ability to revolutionize the healthcare sector. Quantum cryptography technology utilizes the principles of quantum mechanics to secure sensitive information. This makes systems secure against a variety of cyber-attacks [46].

Quantum Key Distribution One of the key uses of quantum cryptography in the healthcare sector is Quantum Key Distribution (QKD). QKD is based on the principles of quantum mechanics This means that if there is an attempt to intercept on the transfer of quantum keys will alter the keys quantum state This change is detectable, immediately alerting the involved parties that the communication has been compromised.

For instance, in telemedicine, where consultations between patients and doctors are increasingly carried out over digital platforms, QKD plays a crucial role. It ensures that all communications and medical file transmissions remain completely secure. This security is vital for maintaining patient confidence and protecting sensitive personal health information from increasingly sophisticated interception attempts.

Protection Against Quantum Computing Threats. The rapid advancement of quantum computing brings a unique challenge. While it can solve complex problems that are currently traditional computers cannot, it also poses a significant risk to standard encryption methods [47]. Quantum computers with their ability to perform calculations at incredibly fast speeds, could decrypt many of today's encryption algorithms, effectively rendering traditional cybersecurity measures obsolete.

However, the field of quantum cryptography is responding to this threat by developing quantum-resistant algorithms. These new algorithms are designed to be secure against both conventional and quantum computers. This ensures that the encryption of sensitive health data remains strong and unbreakable. Implementing these quantum-resistant algorithms in healthcare IT systems is essential for protecting patient data against future quantum attacks, thereby safeguarding privacy and data security in the long term.

Securing Network Infrastructure with Quantum Cryptography Healthcare systems are increasingly relying on digital technologies and internet-connected devices such as electronic health records to IoT-enabled medical devices. Securing the underlying network infrastructure becomes crucial. Quantum cryptography offers solutions here as well, notably through enhanced security protocols for data in transit. For instance, hospitals and clinics that utilize several connected devices for patient monitoring and diagnostics can use quantum cryptography to secure the communication channels within their networks. This security is crucial not only for protecting the integrity of medical data but also for ensuring that operational commands sent to medical devices are not intercepted or tampered with.

7 Navigating Regulatory Compliance and Data Security

Understanding and following regulatory standards is important in today's world. This is especially applicable in regulated areas such as finance and healthcare. Laws such as the HIPAA in the US have been passed and the GDPR of the European Union to protect personal data by setting standards for privacy and security.

7.1 Understanding of HIPAA, GDPR, and Other Relevant Regulations

HIPAA HIPAA is a U.S. laws designed to protect patient health information. It ensures that health care providers, insurance companies, and related organizations handle this sensitive information in a confidential manner. HIPAA outlines specific rules and requirements for maintaining the privacy and security of identifiable health information [48]. The importance of the following has been emphasized here:

Physical Security Secure locations for records.

Operational Security Policies and procedures for managing data processing.

Technological Security Encryption and secure communication channels.

These measures are necessary to prevent unauthorized access and disclosure of patient information.

GDPR The General Data Protection Regulation (GDPR) influences any organization that handles the information of European Union (EU) residents, irrespective of in which the corporation is located. It emphasizes the safety of private records and gives people greater manipulation over their facts [49]. Key points of GDPR encompass:

Strict Consent is Necessary Organizations need to attain clear and express consent from individuals earlier than accumulating or processing their data.

Data Challenge Rights Individuals have rights including getting access to their records, correcting inaccuracies, and soliciting for the deletion in their facts.

Data Processing and Transfer GDPR regulates how records may be processed and moved throughout borders to make sure it stays covered.

Other Laws Different international locations have their personal facts safety legal guidelines tailored to defend their citizens' information. For example:

Canada's PIPEDA (Personal Information Protection and Electronic Documents Act) Sets guidelines for a way groups need to deal with personal records in the route of industrial activities.

Australia's Privacy Act Regulates the coping with of private data by way of each authority and private quarter corporations.

Each of those legal guidelines comes with requirements and emphasizes protecting citizen records from misuse.

7.2 Impact of Regulations on Encryption Practices in Healthcare

Regulations such as HIPAA in the United States and GDPR in Europe significantly impact how organizations handle data encryption:

Encryption Requirements HIPAA and GDPR strongly require or suggest that sensitive information be kept confidential. This is especially important when transmitting data over a network or when storing it. By implementing encryption, organizations add a strong level of security. Without a valid decryption key, it is challenging to decrypt the data. This requirement is an important security measure, protecting everything from personal information to financial reports to health records from unauthorized access [50].

Data Integrity and Privacy Another important aspect of these regulations is to preserve data integrity and confidentiality. Encryption ensures that data is kept confidential so that only authorized users can read it. It also ensures data integrity, which means any unauthorized attempts to modify the data are easily detected. These measures are necessary to prevent data breaches and leaks, which carry significant fines, reputational damage and legal issues [50]. Overall, by requiring encryption, regulations such as HIPAA and GDPR are enforcing stronger security measures, ensuring that the sensitive information is protected. This makes personal data safer and reduces the risk of becoming vulnerable to malicious activity.

8 Advancements and Challenges in Healthcare Data Security

8.1 Advancements

Recent advancements in encryption and data protection are significantly shaping the healthcare sector. Homomorphic encryption now allows secure computations on encrypted data without needing decryption, proving particularly beneficial for secure data sharing and analysis. Blockchain technology offers a decentralized, immutable ledger system that enhances the security and integrity of patient records, increasingly being used in health information exchanges. Quantum cryptography is emerging as a defense against the potential threats posed by quantum computing, with Quantum Key Distribution (QKD) providing ultra-secure communication channels. Additionally, AI and machine learning are playing a growing role in automating threat detection and optimizing encryption processes, enhancing the overall security of healthcare data systems.

8.2 Challenges

Key management remains complex, with poor management leading to risks of data breaches or loss. Balancing security with performance is another critical issue, as encryption can slow down vital healthcare applications, particularly in time-sensitive environments. Regulatory compliance also presents challenges, as organizations must navigate frameworks like HIPAA and GDPR without compromising operational efficiency. The scalability of encryption solutions is increasingly difficult as data volumes grow, requiring methods that maintain security without degrading performance. Lastly, the balance between data accessibility and privacy is delicate, with healthcare providers needing quick access to encrypted patient records while ensuring strong privacy protections.

9 Overcoming Challenges in Implementing Healthcare Encryption

9.1 Key Management and Security

Implementing encryption not only requires the use of strong encryption techniques but also includes maintaining encryption keys [51]. For example, a healthcare facility may use encrypted patient records to protect sensitive information. However, if encryption keys are not properly managed, there is a risk of intrusion and data breaches. To mitigate such risks, healthcare providers or facilities can implement hardware security modules (HSMs) and regularly key rotation and secure the key storage.

9.2 Balancing Security, Privacy and Accessibility

Balancing security with accessibility and utility is the key criteria in determining the efficiency of encryption. The secured and encrypted data need to be available to the authorized systems and users within acceptable time limits. The performance delays due to overengineered encryption mechanisms could improve the data security but could have impact on the accessibility. An approach of role-based access controls and ensure that encryption methods are optimized to allow quick access without compromising security. Apart from this the anonymized data that's utilized for the purpose of research had to maintain the balance between privacy and utility to ensure optimal results. This ensures that the analysis performed on the anonymized data does result with acceptable deviation.

9.3 Ethical Implications of Encryption and Patient Privacy

Encryption is crucial for safeguarding patient privacy, but it comes with its own set of ethical challenges. For example, healthcare utility that encrypts patient information to improve privacy and blocks unauthorized access. While this is a necessary measure, patients often don't fully understand how their data is being protected. Ethical considerations include ensuring that patients understand how their data is protected and obtaining their informed consent. Healthcare providers can address these issues by being transparent about their encryption practices and ensuring that patients are aware of and consent to how their data is handled.

10 Future Trends in Healthcare Data Security and Encryption

10.1 Predictions for the Evolution of Encryption Technologies

As digital technology continues to evolve, so does the landscape of cybersecurity, particularly in the realm of encryption technologies. Here's a deeper look into the technical advancements and trends we can anticipate soon. Encryption technologies are expected to undergo significant transformations to address the increasing complexity of cyber threats. Here are some of the key technical aspects of these advancements:

Adaptive Encryption Future encryption systems will likely employ adaptive encryption techniques that can dynamically adjust encryption strength based on the threat environment. This includes the ability to modify encryption algorithms and key sizes in response to detected threats or vulnerabilities, ensuring optimal security at all times.

Encryption Agility As cryptographic agility becomes crucial, systems will need to support seamless transitions between different cryptographic primitives without service disruption. This agility will help maintain system security, especially when certain encryption algorithms become compromised.

Homomorphic Encryption Enhancements As discussed earlier, homomorphic encryption allows computations on encrypted data. Future developments are expected to focus on enhancing the efficiency and practicality of these algorithms to support more complex and larger-scale computations, making them viable for everyday applications like cloud computing and big data analytics.

10.2 Role of AI and Machine Learning in Healthcare Data Protection

The integration of AI and machine learning into encryption and broader data protection strategies is rapidly advancing. These technologies bring several technical enhancements to the field:

Automated Threat Detection Machine learning algorithms are becoming adept at analyzing patterns in large datasets to identify anomalies that may indicate potential threats or breaches, allowing for rapid response to emerging threats.

Intelligent Encryption Management AI can streamline the management of encryption keys, which includes automated generation, rotation, and revocation of keys. AI can also optimize the deployment of encryption based on data sensitivity and usage patterns, ensuring efficient utilization of resources.

Predictive Security Postures AI models can predict potential security incidents by learning from historical data. This proactive approach in using AI for predictive analytics helps organizations fortify their defenses before breaches occur.

10.3 Preparing for Quantum Computing Challenges

The advent of quantum computing poses unprecedented challenges to traditional encryption methods:

Quantum-Resistant Algorithms The cybersecurity community is actively researching and developing quantum-resistant or post-quantum cryptographic algorithms. These algorithms are based on mathematical problems believed to be resistant to quantum computers, such as lattice-based cryptography, multivariate polynomial cryptography, and hash-based cryptography.

Hybrid Encryption Systems As we transition to a quantum computing era, hybrid systems that combine current cryptographic techniques with quantum-resistant algorithms are being developed. These systems are designed to provide robust security during the period when quantum computers are not yet fully operational but pose a theoretical risk.

11 Conclusion

This chapter has provided a detailed exploration of data encryption and protection techniques essential for safeguarding healthcare data. The rapidly evolving healthcare industry presents both opportunities and challenges in ensuring data security. As technological advancements continue to transform patient care, diagnosis, and treatment methods, the importance of robust data protection mechanisms cannot be overstated. The chapter discusses various encryption strategies, including symmetric and asymmetric encryption, and their applications in protecting data at rest and in transit. Furthermore, emerging technologies such as homomorphic encryption, blockchain, and quantum cryptography were explored for their potential to revolutionize data security practices in healthcare.

Looking ahead, the future of encryption technologies promises significant advancements. Adaptive and agile encryption systems enhanced homomorphic encryption, and the integration of AI and machine learning for automated threat detection and intelligent encryption management are expected to shape the landscape of data security. Additionally, preparing for the quantum computing era by developing quantum-resistant algorithms and hybrid encryption systems will be crucial in maintaining robust data protection. In summary, the continuous evolution of encryption technologies and the proactive adoption of innovative solutions are vital for safeguarding healthcare data. As the industry progresses, ensuring the confidentiality, integrity, and availability of healthcare information will remain a top priority, ultimately contributing to improved patient trust and care outcomes.

References

1. Schlackl, F., Link, N., Hoehle, H.: Antecedents and consequences of data breaches: a systematic review. Inf. Manag. **59**(4) (2022)
2. Song, Z., Wang, G.A., Fan, W.: Firm actions toward data breach incidents and firm equity value: an empirical study. In: Proceedings of the 50th Hawaii International Conference on System Sciences, pp. 4957–4966 (2017)
3. Kolevski, D., Michael, K., Abbas, R., Freeman, M.: Cloud data breach disclosures: the consumer and their personally identifiable information (PII)? In: Proceedings of IEEE Conference on Norbert Wiener 21st Century (CW), pp. 1–9, Jul. 2021

4. Agrafiotis, I., Nurse, J.R.C., Goldsmith, M., Creese, S., Upton, D.: A taxonomy of cyber-harms: defining the impacts of cyber-attacks and understanding how they propagate. J. Cybersecur. **4**(1) (2018)
5. Pandey, A.K., et al.: Key issues in healthcare data integrity: analysis and recommendations. IEEE Access **8**, 40612–40628 (2020). https://doi.org/10.1109/ACCESS.2020.2976687
6. Yang, P., Xiong, N., Ren, J.: Data security and privacy protection for cloud storage: a survey. IEEE Access **8**, 131723–131740 (2020). https://doi.org/10.1109/ACCESS.2020.3009876
7. Pradhan, B., et al.: A comparative study on cipher text policy attribute based encryption schemes. Int. J. Eng. Res. Technol 10 (2021)
8. Gadhiya, N., Tailor, S., Degadwala, S.: A review on different level data encryption through a compression techniques. In: 2024 International Conference on Inventive Computation Technologies (ICICT), Lalitpur, Nepal, 2024, pp. 1378–1381. https://doi.org/10.1109/ICICT60155.2024.10544803
9. Hoener, P.M.: Cloud computing security requirements and solutions: a systematic literature review (2013)
10. Kuzminykh, I., Yevdokymenko, M., Ageyev, D.: Analysis of encryption key management systems: strengths, weaknesses, opportunities, threats. In: 2020 IEEE International Conference on Problems of Infocommunications. Science and Technology (PIC S&T), Kharkiv, Ukraine, 2020, pp. 515–520. https://doi.org/10.1109/PICST51311.2020.9467909
11. Hu, V.C., Ferraiolo, D., Richard Kuhn, D.: Assessment of Access Control Systems. US Department of Commerce, National Institute of Standards and Technology, Gaithersburg, MD (2006)
12. Agrawal, M., Mishra, P.: A comparative survey on symmetric key encryption techniques. Int. J. Comput. Sci. Eng. **4**(5), 877 (2012)
13. Mahadik, S.S., et al.: Digital privacy in healthcare: state-of-the-art and future vision. IEEE Access **12**, 84273–84291 (2024). https://doi.org/10.1109/ACCESS.2024.3410035
14. Yassein, M.B., et al.: Comprehensive study of symmetric key and asymmetric key encryption algorithms. In: 2017 International Conference on Engineering and Technology (ICET). IEEE (2017)
15. Gupta, K., Silakari, S.: Ecc over rsa for asymmetric encryption: a review. Int. J. Comput. Sci. Issues (IJCSI) **8**(3), 370 (2011)
16. Singh, G.: A study of encryption algorithms (RSA, DES, 3DES and AES) for information security. Int. J. Comput. Appl. **67**(19) (2013)
17. Ulybyshev, D., et al.: Protecting electronic health records in transit and at rest. In: 2020 IEEE 33rd International Symposium on Computer-Based Medical Systems (CBMS). IEEE (2020)
18. Khati, L., Mouha, N., Vergnaud, D.: Full disk encryption: bridging theory and practice. In: Topics in Cryptology–CT-RSA 2017: The Cryptographers' Track at the RSA Conference 2017, San Francisco, CA, USA, February 14–17, 2017, Proceedings. Springer (2017)
19. Balogun, A.M., Zhu, S.Y.: Privacy impacts of data encryption on the efficiency of digital forensics technology (2013). arXiv:1312.3183
20. Miao, Q.: Research and analysis on encryption principle of TrueCrypt software system. In: 2nd International Conference on Information Science & Engineering, pp. 1409–1412 (2010)
21. Carrier, B.: File System Forensic Analysis. Addison-Wesley, Upper Saddle River, NJ (2005)
22. TrueCrypt Foundation: TrueCrypt Documentation. Available at http://www.truecrypt.org/docs/ Retrieved 15 Feb 2013 (2012)
23. E-security Planet: Buyer's Guide to Full Disk Encryption. Available at http://www.esecurityplanet.com/mobile-security/buyersguide-to-full-disk-encryption.html Retrieved 18 Jan 2013 (2012)
24. Bouganim, L., Guo, Y.: Database encryption. In: Encyclopedia of Cryptography and Security, pp. 1–9 (2009)
25. Mattsson, U.T.: Database encryption-how to balance security with performance. Available at SSRN 670561 (2005)
26. Joshi, M., Joshi, M.: A review on various data encryption techniques used in cloud enabled data centre. In: 2020 International Conference on Intelligent Engineering and Management (ICIEM). IEEE (2020)

27. Chandramouli, R., Iorga, M., Chokhani, S.: Cryptographic key management issues and challenges in cloud services. In: Secure Cloud Computing, pp. 1–30 (2013)
28. Daoud, L., Huen, H.: Performance study of software-based encrypting data at rest. In: Proceedings of 37th International Conference, vol. 82, pp. 122–130 (2022)
29. Meier, R.: Data in transit security. In: Trends in Data Protection and Encryption Technologies, pp. 135 (2023)
30. Yigzaw, K.Y., et al.: Health data security and privacy: challenges and solutions for the future. In: Roadmap to Successful Digital Health Ecosystems, pp. 335–362 (2022)
31. Saxena, A., Kaulgud, V., Sharma, V.: Application layer encryption for cloud. In: 2015 Asia–Pacific Software Engineering Conference (APSEC). IEEE (2015)
32. Frankel, S., et al.: Guide to IPsec VPNs, pp. 800–77 (2005)
33. Blobel, B., et al.: 9 Standard guide for implementing HL7 communication security. In: Security Standards for Healthcare Information Systems, pp. 183–219. IOS Press (2002)
34. Zhang, J.: DICOM image secure communication with internet protocols. Teleradiology 33–47 (2008)
35. Ecarot, T., et al.: Sensitive data exchange protocol suite for healthcare. In: 2020 IEEE Symposium on Computers and Communications (ISCC). IEEE (2020)
36. Edo, O.C., et al. A zero trust architecture for health information systems. Health Technol. **14**(1), 189–199
37. Kaur, K., Gupta, I., Singh, A.K.: A comparative evaluation of data leakage/loss prevention systems (DLPS). In: Proceedings of 4th International Conference on Computer Science & Information Technology (CS & IT-CSCP) (2017)
38. Prabhakar, R., et al.: Securing disk-resident data through application level encryption. In: Fourth International IEEE Security in Storage Workshop. IEEE (2007)
39. Atadoga, A., et al.: A comparative review of data encryption methods in the USA and Europe. Comput. Sci. IT Res. J. **5**(2), 447–460 (2024)
40. Bameyi, O.J., et al.: End-to-end security in communication networks: a review. In: Innovations in Bio-Inspired Computing and Applications: Proceedings of the 11th International Conference on Innovations in Bio-Inspired Computing and Applications (IBICA 2020) Held During 16–18 Dec 2020 11. Springer (2021)
41. Rocha, V.F., López, J., Falcão Da Rocha, V.: An overview on homomorphic encryption algorithms. In: Technical Report. UNICAMP Universidade Estadual de Campinas (2018)
42. Acar, A., et al.: A survey on homomorphic encryption schemes: theory and implementation. ACM Comput. Surveys (CSUR) **51**(4), 1–35 (2018)
43. Kogos, K.G., Filippova, K.S., Epishkina, A.V.: Fully homomorphic encryption schemes: the state of the art. In: 2017 IEEE Conference of Russian Young Researchers in Electrical and Electronic Engineering (EIConRus). IEEE (2017)
44. Agbo, C.C., Mahmoud, Q.H., Mikael Eklund, J.: Blockchain technology in healthcare: a systematic review. Healthcare **7**(2) (2019) (MDPI)
45. Zheng, Z., et al.: An overview of blockchain technology: architecture, consensus, and future trends. In: 2017 IEEE international Congress on Big Data (BigData Congress). IEEE (2017)
46. Portmann, C., Renner, R.: Security in quantum cryptography. Rev. Mod. Phys. **94**(2), 025008 (2022)
47. Faruk, Md J.H., et al.: A review of quantum cybersecurity: threats, risks and opportunities. In: 2022 1st International Conference on AI in Cybersecurity (ICAIC). IEEE (2022)
48. Stevens, G.M., American Law Division: A brief summary of the HIPAA medical privacy rule. In: Congressional Research Service, the Library of Congress (2003)
49. Li, H., Lu, Y., He, W.: The impact of GDPR on global technology development. J. Glob. Inf. Technol. Manag. **22**(1), 1–6 (2019)
50. Stallings W.: Information Privacy Engineering and Privacy by Design: Understanding Privacy Threats, Technology, and Regulations Based on Standards and Best Practices. Addison-Wesley Professional (2019)
51. Barker E., Barker, W.: Recommendation for key management, Part 2: best practices for key management organization. In: No. NIST Special Publication (SP) 800-57 Part 2 Rev. 1 (Draft). National Institute of Standards and Technology (2018)

Venkata Nedunoori an accomplished technology leader and IEEE Senior Member, brings over 16 years of expertise across sectors such as Insurance, Securities, Airlines, Media, and Advertising. He specializes in designing and implementing sophisticated cloud-based solutions, with a focus on building scalable, secure, and cost-effective platforms that enable organizations to excel in a rapidly evolving digital environment. With a strong passion for cloud security and artificial intelligence, Venkata is dedicated to enhancing digital ecosystems and driving innovation as technology advances.

Artificial Intelligence (AI) Adoption and Implementation Challenges in Indian Healthcare Sector

Madhavi Kilaru and Rajasekhara Mouly Potluri

Abstract Artificial intelligence (AI) has become a critical and foundational component across various fields, significantly enhancing efficiency, decision-making, and driving innovation. By automating routine tasks, analyzing vast datasets for insights, personalizing consumer experiences, and optimizing operations, AI delivers substantial benefits such as cost savings, improved outcomes, and more effective responses to complex challenges. In a country like India, with its large population and high demand for healthcare services, AI holds immense potential to address the shortage of healthcare professionals and enhance the overall healthcare system. The adoption of AI in healthcare is essential to streamline administrative tasks, allowing healthcare professionals to focus more on providing direct patient care. While AI is already making strides in multiple sectors across India, the healthcare industry still requires broader and more impactful AI applications to improve access to care and clinical outcomes. AI can play a transformative role in enhancing diagnostic accuracy, optimizing the allocation of resources, and streamlining patient management, which are crucial to meeting the healthcare demands of India's vast population. However, the implementation of AI in Indian healthcare faces several challenges, including infrastructural limitations, data privacy concerns, and the need for upskilling the workforce to effectively use AI technologies. This chapter focuses on the adoption of AI in the Indian healthcare sector, exploring both the opportunities it offers and the obstacles that must be addressed to ensure successful integration. This chapter examines how AI can assist in overcoming critical challenges and help the healthcare system operate more efficiently, ultimately improving patient outcomes and healthcare access across India.

Keywords Artificial intelligence · Healthcare innovation · Diagnostic accuracy · Resource optimization · Patient management · AI adoption in India · Healthcare access · Workforce upskilling · Data privacy · Healthcare infrastructure

---

M. Kilaru
VNR Vignana Jyothi Institute of Engineering and Technology, Hyderabad, India

R. M. Potluri (✉)
Kazakh-British Technical University, Almaty, Kazakhstan
e-mail: prmouly24@gmail.com

N. R. Vajjhala et al. (eds.), *Artificial Intelligence in Healthcare Information Systems—Security and Privacy Challenges*, Information Systems Engineering and Management 34, https://doi.org/10.1007/978-3-031-84404-1_9

1 Introduction

Artificial Intelligence (AI) significantly impacts changing areas such as education, supply chain management, and finance. AI now encompasses practically all key sectors and assists people in doing innovative work rather than hard work. It is functioning in a much more targeted manner worldwide in many fields. It can accomplish activities that need conventional human intellectual knowledge, such as comprehending data patterns, language understanding, making sound decisions, and gaining and learning from experience by constructing computer systems [1]. Recently, one of the crucial and rising sectors, healthcare AI, has played a significant role in demonstrating the potential and exploiting the skills to enhance the healthcare system through patient care, test diagnosis, and various administrative tasks. AI is already essential in healthcare; it is not a futuristic vision. AI applications have been successfully implemented in the present days, demonstrating the potential growth in healthcare data sets, advancement in computer applications, and significant breakthroughs in machine learning [2]. The value of AI applications in healthcare cannot be emphasized. AI can transform disease diagnosis, tailor treatment to specific patients, monitor a patient's health status in real-time, and manage various administrative and operational delivery. For example, AI applications in diagnoses such as medical pictures may be overlooked by humans due to the vast data sets. This precision leads to earlier and more accurate diagnoses, significantly impacting patient outcomes. Similarly, AI algorithms may sift through large datasets in patient customization treatment to detect trends and anticipate which medicines will be most beneficial for individual patient profiles, ushering in genuinely personalized medicine [3].

Furthermore, AI applications extend to patient monitoring, including wearable gadgets and remote monitoring systems that offer continuous oversight of patient health, enabling timely interventions and reducing hospital readmissions. In terms of healthcare delivery, AI can streamline operations, from scheduling appointments to optimizing hospital workflows, thereby improving efficiency and patient satisfaction. While many healthcare centers are integrating AI applications in their companies, AI adoption is still in its early stages, notably in clinical interventions [4]. Most of the highlighted use cases are still in the development and testing phase. Most existing use cases involve decision support systems, process optimization, and virtual assistants. Computer vision, one of AI's more advanced uses, is being used to train AI algorithms to read X-rays and scans to help with disease identification and diagnosis. Only a few companies produce surgical simulations, individualized health solutions, and patient monitoring systems [5]. Furthermore, a tiny number of interventions use NLP and speech recognition, both of which are critical for meeting diverse linguistic, and literacy needs in our country. That is why the book chapter focuses solely on how real-time AI algorithms are being used in the Indian healthcare system, as well as the problems encountered during the deployment process.

2 Current State of AI Adoption in Healthcare

AI logarithms play an essential role in virtual assistants, and catboats have made their way into healthcare settings, giving patients individualized treatment even in remote locations while delivering a wealth of information. These intelligent systems can answer medical queries, provide correct and practical self-care advice, and provide appropriate information and Medicare based on the patient's symptoms. This increases access to healthcare, minimizes the pressure on providers, and has proven helpful in predictive analytics and patient monitoring. By examining massive volumes of patient data, AI systems can uncover illness patterns and risk factors, allowing healthcare providers to react early and avert unpleasant occurrences. AI-powered wearable gadgets and remote monitoring systems allow continuous monitoring of vital signs, providing real-time alerts for critical changes in a patient's health status. Applications in healthcare in India are rapidly evolving and encompass various domains, significantly improving patient care, operational efficiency, and diagnostic accuracy [6].

2.1 Diagnostic Assistance

AI algorithms can help evaluate medical pictures (such as X-rays, MRIs, and CT scans) to identify anomalies, such as tumors or fractures, with high accuracy. AI systems can spot patterns in patient data, assisting in disease diagnosis based on symptoms, lab results, or medical history AI solutions give clinicians with recommendations or alerts, allowing them to make better-informed diagnosis and treatment plans. This includes businesses, laboratories, and other investigative and diagnostic institutions. Besides Google and IBM, India has seen many newly founded companies focused on using AI to diagnose diseases and conditions [7]. Diagnostics also uses descriptive and predictive AI in India. Artelus, for example, works with deep learning algorithms to develop expert screening gears for medical practitioners and clinics to diagnose diseases such as diabetes, tuberculosis, breast cancer, and many others. It also provides intelligent early diabetes detection through its advanced AI-based algorithm, which can be done with a single eye check. Artelus has achieved maximum penetration in rural India by collaborating with Shushruth Hospital to provide Diabetic Retinopathy clinics in villages where medical facilities are understaffed [8]. Artelus has brought Diabetic Retinopathy screening to remote corners of the world by utilizing AI on a-chip technology. Their offline service also delivers point-of-care solutions to underprivileged communities that do not require an internet connection. Every patient's data is coded with numbers to protect patients' privacy.

Healthcare datasets, such as literature and case studies, were put into the machine for machine learning. When a computer learns, all practical instances are tested, which is known as normative learning, which is faster than initial learning and allows the system to screen the most significant number of diabetics. Breast cancer is the most common cause of death among women today. According to the WHO report, one in every twelve women is at risk of developing breast illness. NIRMAI health analytics makes use of thermal analytics, a unique artificial intelligence-based medical gadget that detects breast cancer sooner than standard approaches. This is fully radiation-free, painless, and can help women of all ages. This provides accurate solutions for identifying breast cancer. It also serves all types of people. Advenio is a deep-tech healthcare start-up and social enterprise that employs artificial intelligence, data science, and advanced image processing to revolutionize healthcare diagnosis. Advenio's patented AI-based algorithms are intended to give precise differential diagnoses, culminating in an AI system that helps clinicians make clinical decisions. Chest X-rays can assist in diagnosing tuberculosis, whereas ultrasound images can help detect serious infections. Orbuculum predicts cancer, diabetes, and other chronic diseases using artificial intelligence and genetic data. Orbuculum's artificial intelligence technology will assist clinicians in diagnosing chronic diseases earlier, lowering costs and increasing efficacy [9]. Cureskin detects nearly six types of common skin ailments, including pimples, acne, scars, dark spots, pigmentation, and dark circles, and recommends treatment via mobile apps. As time passes, the dreadful job and survival struggles that people face in their lives lead to mental health issues.

Recovery from all these disorders is an arduous effort for doctors and requires patience. In order to make the world more mentally resilient, we created Wysa, an AI chatbot that uses evidence-based cognitive-behavioral approaches (CBT) to help you feel heard. Wysa and expert human support deliver high-quality mental health care 24 h a day, seven days a week. Wysa is a top mental health app that uses AI for counseling and therapy. Wysa uses an AI-powered chatbot in its applications, allowing users to engage in emotionally stimulating discussions that are therapeutic for anxiety, despair, and stress [10]. Other alternative services include Woebot, Able To, Talkspace, Sanvello, Headspace, Betterhelp, Mindshift CBT, and Moodfit are some of the alternative options available for dealing with emotional intelligence concerns.

2.2 *Drug Discovery and Development*

Drug discovery is an essential challenge for service providers because new viruses are introduced into the environment daily. Finding the right medicine is a great struggle for service developers and involves risk because they provide survival to people worldwide. Historically, drug discovery has been a challenging, time-consuming, expensive, and frequently frustrating task. Bringing a new treatment to market often takes over a decade and billions of dollars, with clinical studies failing at a stunning 90% rate. This inefficiency stresses healthcare institutions and delays the delivery of essential medications to individuals in need. Enter AI-powered drug discovery by leveraging advanced machine learning techniques, natural language processing, and predictive modeling. AI is reshaping every stage of the pharmaceutical R&D pipeline.

Using artificial intelligence (AI) in medication research and development is transforming the pharmaceutical industry in India. AI technologies allow researchers to examine large datasets quickly, find possible drug candidates, and anticipate chemical interactions using machine learning models. Generative AI creates unique chemical designs and optimizes lead molecules efficiently. Furthermore, AI improves clinical trial operations by speeding up patient recruitment and optimizing study protocols, resulting in faster timelines and better outcomes. Another captivating application is drug repurposing, which uses AI to identify current drugs that can be used for new therapeutic applications, considerably reducing development time and costs [11].

AI is changing drug discovery and development in India through novel real-time applications. Companies such as Biocon and CureMetrix use AI to evaluate large biological datasets, which improves the identification of possible medication candidates. For example, Biocon uses machine intelligence to optimize lead chemicals, increasing efficacy while decreasing development time. Biocon offers a whole range of insulins. Niramai uses artificial intelligence to repurpose pharmaceuticals, identifying existing treatments that may be beneficial against novel diseases such as COVID-19. This method saves time and lowers the expenses of bringing innovative therapeutics to market [12]. Start-ups like SigTuple use AI to automate pathology diagnosis, giving significant insights for drug efficacy research. These developments are supported by programs from the Biotechnology Industry Research Assistance Council (BIRAC), reflecting a robust ecosystem.

2.3 Telemedicine and Virtual Care

Telemedicine can significantly cut healthcare expenditures, enhance patient reach and health outcomes, and alter how healthcare providers serve patients. This scope introduces people to telemedicine. Observing people's hectic schedules and other factors indicates that the future of health care will be hybrid. In other words, we may anticipate that patients will increasingly receive a combination of virtual and in-person therapy in the future—a high level. Medical app downloads have surged by 60% in the last year. As a result, virtual medicine, or the face-to-face hybridization of medical care, holds the key to the future of telemedicine. Telemedicine had a significant role during the Covid epidemic, and a review was done to find out how telemedicine affected the community during this period. Telemedicine played a crucial role during the Covid epidemic, and an assessment was done to determine how it influenced the population. Overall, improvements in telemedicine are determined by the circumstances at hand. Although telemedicine and telehealth may still improve in the future as technology advances, without the pandemic, growth or progress may be [13]. The World Health Organization advises a meager doctor-to-population ratio, which must increase to meet India's enormous population. It will take time and money to find new, qualified doctors, so it seems likely that the doctor-to-patient ratio will stay low for some time. The active telemedicine services around the nation partially help to offset this shortfall.

The Ministry of Health and Family Welfare and the Department of Information Technology jointly oversee these national services. In order to carry out a green field project on e-health, the MoHFW, GOI Telemedicine division has established a National Telemedicine Portal. This includes creating a National Rural Telemedicine Network to deliver e-healthcare and a National Medical College Network (NMCN) to connect medical colleges nationwide. The National Digital Health Authority of India (NDHAI)/National e-health authority (NeHA) is being established as a cornerstone of the e-health wing of the National Health Portal (NHP) to deliver high-quality health services for all citizens through the cost-minimization and secure use of virtual tools in health and health-related fields. MoHFW created a set of Electronic Health Records (EHR) standards in 2013 and a revised version of the same in 2016 to ensure the safe transmission of information during telemedicine procedures. Traditional medicine is one area in which telemedicine is used in India. Through telemedicine, the National Rural AYUSH Telemedicine Network seeks to increase the population's awareness of the advantages of conventional medical practices. ISRO developed the concept of village resource centers (VRCs), which operate based on satellite communication, in order to increase a wide range of services, primarily based on the rural population, such as telemedicine, tele-education, online decision support, interactive farmers' advisory services, tele fishery, e-governance, weather services, and water management. The local Resource Centers (VRCs) function as educational spaces and connect to specialty hospitals, allowing the villages to offer highly skilled medical specialists to the local population to attract their interest. There are already close to 500 of these VRCs operating all over the country. Another online mobile telemedicine conglomerate that unifies numerous hospitals, mobile clinics, medical professionals, and rural mobile units is called AROGYASREE. The Indian Council of Medical Research (ICMR) is the organization behind the project. They are working with a group of German scientists at the University of Karlsruhe to create an ECG jacket that patients can wear to continuously monitor their ECG without being admitted to the hospital [14].

2.4 Predictive Analytics

Predictive analytics is used in the healthcare industry to aid in analyzing current and historical healthcare data. It allows healthcare professionals to identify new opportunities to make more effective and efficient operational and clinical decisions, forecast data trends, and even help manage disease spread. Using the example of the COVID-19 pandemic, BlueDot in Canada developed predictive analytical methods that provided early warning of illness spread in Wuhan. Another tool designed for disease tracking at the University of Texas Health Science Center at Houston (UTHealth) aided during the pandemic by indicating the disease's current and future trends. During the pandemic, analytics played an essential and influential role in India. For example, India achieved success in the implementation of AI technology in the pandemic response system, including diagnosis, clinical decision-making, patient management, clinical services, and facial recognition with masks. AI also assisted in other industries during the pandemic, including BFSI, e-commerce, sales, and marketing, in the transition from contact-based to contactless services during the country's lockdowns.

During the pandemic, AI in India surged by 70% in 2020. Across the country, pandemic vaccines such as CoWIN are administered safely and cost-effectively with effective protocols [15]. Haptik, a conversational AI firm, created the Corona Helpdesk to raise awareness nationwide. My Gov Saathi's chat board provides accurate and helpful information about illness spread. Saathi works only on data analytics and natural language processing to provide daily services to users. Data processing is used in the main category of predictive analysis. Data mining is primarily used in healthcare to train medical personnel in making clinical decisions by anticipating illnesses and offering diagnostic support. Data mining is the process of identifying patterns in massive data sets and using that information to create prediction models.

Many sectors make good use of information mining. It helps the client to respond to the retail sector model. It allows banks to predict the profitability of their customers. It supports medium, manufacturing, automotive, higher education, life sciences, and other related use cases. However, information processing remains focused, with only a few practical success stories for primary instructional practice. Academics use data mining approaches to publish studies like decision trees, clusters, neural networks, and time series. Healthcare professionals can diagnose, treat, and prevent people with physical and mental problems in several ways. In most countries, the healthcare industry is growing at a dizzying pace. Because of their bulk, medical records, administrative reports, and other benchmarking findings produced by the healthcare industry are regarded as data-rich. However, these medical records are underutilized. Large amounts of data can be mined to reveal new and essential information. Data mining in healthcare is mainly used to help clinicians make clinical decisions by anticipating illnesses and giving diagnostics.

As organizations need to acquire larger data sets with needed insights into business activities, recognizing abnormal data or outliers is vital in discovering inefficiencies, rare events, the core cause of difficulties, or chances for operational changes. It identifies the most significant changes to data collection. Critical diagnostic approaches such as standard support vector data description, density-induced support vector data description, and Gaussian combination are used to evaluate anomaly detection accuracy on an uncertain dataset. Clustering is a common expressive task that includes identifying a small number of categories or clusters to describe the data. The clustering method used the vector quantization technique to predict readmissions in advanced medicine. Classification is discovering a learning prediction characteristic that assigns a data item to one of the multiple specified classes; the Mahalanobis Taguchi system (MTS) technique was utilized for multivariable statistical assessment. The Mahalanobis distance (MD) compares the abnormality of two clusters, whereas the Mahalanobis room (MS) compares the abnormality of data from a known reference group. Pressure ulcers were predicted using the MTS. Class imbalance is prevalent in healthcare databases.

When dealing with skewed or unbalanced data sets, skewed distribution frequently impacts data mining methods. This challenge usually leads to very exact classification accuracy for the majority class and low precision for the minority class. This approach is ideal for data analysis since it can show the degree of identified abnormality. This strategy is also used to scale the MD. This method's performance is based on the vast disparity between typical and atypical instances. During the test phase, the MTS shows better sensitivity and scaling. The KNN (K Nearest Neighbors) algorithm evaluates and categorizes all available data points before classifying new cases based on the established categories. It helps identify patterns and make estimates.

One of the most essential data mining techniques is KNN. It is a terrific strategy that is simple and straightforward to use in healthcare [16]. It is known as Memory-Based Classification because the training examples must be kept in memory at runtime. When dealing with continuous attributes, the Euclidean distance is used to calculate the difference between them. When using the Euclidean distance formula, one crucial issue is that the frequency of high values overshadows that of smaller ones. For example, among patients seeking heart disease records, the cholesterol level fluctuates between 100 and 190, while the age spans from 40 to 80. As a result, the cholesterol level will have a more significant influence than age. To address this issue, continuous attributes are normalized to have the same impact on the distance measure between instances.

2.5 Remote Patient Monitoring

The vital signs monitor is a crucial piece of equipment that aids the medical staff in taking accurate, real-time patient physiological data readings. It measures blood pressure noninvasively and invasively and analyzes an electrocardiogram (ECG), heart rate, pulse oxygen saturation, respiratory rate, and temperature. These monitors can be modified into wearable gadgets if artificial intelligence is incorporated into the process. The AI-integrated vital signs monitor that was only recently released by VitaSensors is a prime example, displaying its capability to understand patient data and convey the information to physicians for fast decision-making.

Physical activity monitoring tracks patients' daily activities and categorizes them based on intensity or body position, with AI playing a significant role. This monitoring technique is also helpful for spotting irregularities in the early stages of the disease. These monitors now have access to critical tools that enable them to go beyond the limitations of traditional activity tracking. These AI systems can assess and categorize patients' activities based on their intensity level or body posture and issue emergency alerts. Consider the fall detection function, an AI-enabled functionality that is extensively used by several industry giants, including Google, Apple, Samsung, and others [17].

The RPM technique might vary depending on the individual using the device and how to monitor using the technology, and most technologies have comparable components. The wireless device evolves with specified physiological parameters and stores the data that has been acquired. This storage depends on how it interacts with sensors, health professional databases, and related applications. These programs provide users with an interface for tracking or analyzing data and valuable therapeutic recommendations. It acts as virtual health care, which has grown in popularity and emerged as an essential tool for chronic disease management. RPM has transformed the landscape of chronic disease management by allowing for continuous health tracking outside of the confines of a clinic. When paired with AI, healthcare personnel's ability to manage patient data improves, increasing RPM's chances of discovering cures for chronic diseases. Consider, for example, using the concept of RPM to monitor drug adherence, which is a critical component in managing chronic diseases. In this context, generative AI acts as a virtual companion, sending out thoughtful reminders and messages to boost patient involvement in the process of treatment. Furthermore, improved levels of adherence make it possible to collect data more frequently and with a higher degree of precision, which provides the way for advances in the healthcare sector [18].

3 Barriers and Challenges to AI Adoption in Healthcare

In the age of substantial public digital data repositories, data privacy has become increasingly crucial; otherwise, it leads to improper treatment, impacting individuals' lives. This is especially true in healthcare, where data may be leveraged by tracking individuals directly, opening many options. Healthcare practitioners are typically unsure about the ethical and regulatory constraints that govern data sharing despite being responsible for creating research and goods that may benefit products. These are significant challenges confronting regulators when evaluating artificial intelligence as a diagnostic aid. However, there is one more area that, yet in its infancy, should not be ignored. We have already noted its use in medication development. However, as with other technology, such as nuclear energy, things can be utilized for good or evil. While these algorithms typically screen out toxicity, a group of academics took a different approach. In a proof-of-concept study, they demonstrated that within a few hours, they could create analogs of known chemical weapons that were projected to be much more toxic. They call for increased awareness of the scope of dual usage of artificial intelligence, ethical training for those involved, and channels for reporting potential abuses [19]. This provides another previously primarily ignored challenge for those regulating artificial intelligence that falls on the margins of existing technology assessment models. Regulatory weaknesses, such as the lack of a Regulating Authority for AI India in the healthcare system, threaten the adoption of such issues, as does the necessity for an effective qualification process. Provided it is not limited to a single issue or feature, there is also a need for multiple regulators for specific aspects, such as the Medical Council of India, and for data, a new regulator under the Data Protection Bill. Another concern with AI, particularly neural networks, is the black box. It is programmable, but technicians do not fully understand its internal functions, raising the question of usefulness.

Regulatory authorities are trying to keep up with AI's black box existence, which poses a specific collection of regulatory issues. Although some artificial intelligence systems operate under transparency and are easily understandable, others are unpredictable, thus called black boxes. Regulatory authorities cannot review applications involving a black-box AI/ML approach without clarifying and appreciating how success has been achieved. A further regulatory challenge is that AI's continuous learning mechanism modifies its output in response to the new data provided over time. Identifying black boxes is extremely difficult, and AI algorithms that continuously update themselves raise safety risks that no regulatory agency has addressed. Conventional healthcare legislation cannot be used for AI-based therapy because it is not a medicine or vaccination. AI is concerned with machine learning, and the algorithm continues to evolve due to its 'learning' capabilities. When regulatory approval is given for an algorithm, the algorithm "learns" from new data applied, evolving and becoming a whole different algorithm. AI, like other emerging healthcare technologies, requires governmental approval.

Startups are also expected to produce medical trial proof when introducing their concept to the healthcare industry to grab medical practitioners' interest. Furthermore, there is no clear norm to follow while conducting clinical trials. To address the safety and quality of AI-driven healthcare systems, a proper certification scheme is required, and it can assist in developing confidence between healthcare professionals and patients. Many healthcare providers, particularly in India, are experiencing challenges with infrastructure. Significant investment is required to implement AI techniques in the healthcare industry. Healthcare experts and the government are both neglecting to construct adequate healthcare infrastructure. For example, most cloud computing apps rely on servers provided by foreign corporations. These days, healthcare frameworks adjust to computerized stages and become more patient-centered and data-driven. Low investment in local infrastructure has also contributed to the relocation of many Indian start-ups to other countries, where access to capital and technology is more readily available. Infrastructure difficulties are widespread in developing-world rural locations. Electricity and the Internet are inaccessible to some regions, which must be addressed during planning. Furthermore, India's lack of technical infrastructure leads to limited deep learning capabilities, which is a significant barrier to AI adoption in the healthcare sector. The lack of a substantial native-install cloud-on-demand network in India prevents government-funded research facilities from accessing recent AI advances. Many businesses cannot risk storing their data outside of India and accessing it using algorithms the Indian government has no direct authority. This restriction implements AI technologies, increasing the private sector's cost.

Without this critical infrastructure, India has struggled to achieve the tipping point in AI output and profitability, and it will continue to suffer, even as market adoption accelerates. Many healthcare experts have gone to other countries since they earn more money outside India. The AI efforts in healthcare have attracted much attention, but AI's adoption issues have not been significantly addressed. Its implementation has several challenges, such as resolving the unequal relationship between trained physicians and patients, increasing physicians' efficiency to be more effective in their work, providing AI-enabled healthcare equipment in rural communities, and educating physicians or doctors in handling it. AI technologies have the potential to enhance patient outcomes. Still, they may also pose significant risks in terms of inadequate patient risk assessment, medical error, suggestions for treatment, privacy violations, and others [20].

4 Conclusion

Scaling up and deploying technologies in India is challenging. There is much optimism for AI in healthcare, but progress and advancements in other sectors must be made before AI technology can be used safely and ethically. Society must solve regulatory, privacy, and socio-cultural issues, although some AI software platforms will help alleviate technical and technological debt constraints. The Indian government has launched many steps to address privacy and regulatory concerns. The growing need for technology to address capital limitations in India and AI's potential to give affordable solutions suggests that India may soon be ready to reap the benefits of these technologies regarding healthcare outcomes. The government should give funds and opportunities to the commercial or public sectors to invest in AI. Aligning private enterprises' economic incentives with social responsibility and public health improvement is crucial. Innovative, sustainable, and scalable artificial intelligence technology in India has the potential to enhance health outcomes greatly. Novel, new AI-based solutions are critical to protecting healthcare data privacy.

This chapter investigates various cutting-edge methodologies and opens a fundamentally exciting research area. The theoretical foundation of modern healthcare data privacy protection approaches, and an overview of a paradigm formed by combining three techniques: federated learning, cryptographic approach, and differential privacy. These strategies improve patient data security and privacy while promoting safe, trusted, and efficient care. As a result, healthcare shareholders should use these strategies in their AI training algorithms to create AI models that comply with industry, national, and global data security regulations. In addition, countries and healthcare organizations should actively participate in establishing a worldwide legislative architecture for healthcare data governance in order to create a global network for information governance in the medicine domain and extend data exchange possibilities. Given the uncertainty in information governance regulations between countries, increasing information protection and promoting cross-border information flow via global soft law is critical. Any concept or component developed for healthcare must always obtain approval from the World Health Organization in order to secure crucial data.

References

1. Sarker, I.H.: AI-based modeling: techniques, applications and research issues towards automation, intelligent and smart systems. SN Comput. Sci. **3**(2), 158 (2022)
2. Panesar, A.: Machine Learning and AI for Healthcare, pp. 1–73. Apress, Coventry, UK (2019)
3. Singh, A.P., Saxena, R., Saxena, S., Maurya, N.K.: Artificial intelligence revolution in healthcare: transforming diagnosis, treatment, and patient care. Asian J. Adv. Res. **7**(1), 241–263 (2024)
4. Ali Mohamad, T., Bastone, A., Bernhard, F., Schiavone, F.: How artificial intelligence impacts the competitive position of healthcare organizations. J. Organ. Chang. Manag. **36**(8), 49–70 (2023)
5. Manickam, P., Mariappan, S.A., Murugesan, S.M., Hansda, S., Kaushik, A., Shinde, R., Thipperudraswamy, S.P.: Artificial intelligence (AI) and the Internet of Medical Things (IoMT) assisted biomedical systems for intelligent healthcare. Biosensors **12**(8), 562 (2022)
6. Kalra, N., Verma, P., Verma, S.: Advancements in AI-based healthcare techniques with FOCUS ON diagnostic techniques. Comput. Biol. Med. **179**, 108917 (2024)
7. Kumar, Y., Koul, A., Singla, R., Ijaz, M.F.: Artificial intelligence in disease diagnosis: a systematic literature review, synthesizing framework, and future research agenda. J. Ambient. Intell. Humaniz. Comput. **14**(7), 8459–8486 (2023)
8. Roy, M., Jamwal, M.: An overview of artificial intelligence (AI) intervention in the Indian healthcare system. Int. J. Sci. Res. (IJSR) **9**(8), 1226–1230 (2020)
9. Mukherjee, S., Chittipaka, V., Baral, M.M., Pal, S.K., Rana, S.: Impact of artificial intelligence in the healthcare sector. Artif. Intell. Indus. **4**, 23–54 (2022)
10. Pavlopoulos, A., Rachiotis, T., Maglogiannis, I.: An Overview of Tools and Technologies for Anxiety and Depression Management Using AI (2024)
11. Bhattamisra, S.K., Banerjee, P., Gupta, P., Mayuren, J., Patra, S., Candasamy, M.: Artificial intelligence in pharmaceutical and healthcare research. Big Data and Cognit. Comput. **7**(1), 10 (2023)
12. Chavda, V.P., Patel, K., Patel, S., Apostolopoulos, V.: Artificial intelligence and machine learning in the healthcare sector. In: Bioinformatics Tools for Pharmaceutical Drug Product Development, pp. 285–314. (2023)
13. Kothamali, P.R., Banik, S., Dandyala, S.S.M., Kumar Karne, V.: Advancing telemedicine and healthcare systems with AI and machine learning. Int. J. Mach. Learn. Res. Cybersecurity and Artif. Intell. **15**(1), 177–207 (2024)
14. Padmavati, S.: In: My Life and Medicine. Notion Press (2018)
15. Hasan, I., Jabeen, A., Rizvi, S.A.M.: Next-generation technologies implicated in public health management of COVID-19 pandemic. Int. J. Electron. Healthc. **13**(2), 91–112 (2023)
16. Ogundele, I.O., Popoola, O.L., Oyesola, O.O., Orija, K.T.: A review on data mining in healthcare. Int. J. Adv. Res. Comput. Eng. Technol. (IJARCET) **7**, 698–704 (2018)
17. Alsareii, S.A., Awais, M., Alamri, A.M., AlAsmari, M.Y., Irfan, M., Aslam, N., Raza, M.: Physical activity monitoring and classification using machine learning techniques. Life **12**(8), 1103 (2022)
18. Nayak, A., Satpathy, I., Patnaik, A., Patnaik, B.C.M., Goswami, V.: Remote patient monitoring (RPM) with biosensors: reforming healthcare beyond walls of hospital. In: 2024 International Conference on Advancements in Smart, Secure and Intelligent Computing (ASSIC), January, pp. 01–07. IEEE (2024)
19. Renda, A.: Artificial Intelligence. Ethics, Governance and Policy Challenges. CEPS Centre for European Policy Studies (2019)
20. Nizam, V., Aslekar, A.: Challenges of applying AI in healthcare in India. J. Pharmaceut. Res. Int. **33**(36B), 203–209 (2021)

Madhavi Kilaru , an Assistant Professor in the Department of Humanities & Sciences at VNR Vignana Jyothi Institute of Engineering & Technology, Hyderabad, has about 17 years of teaching experience. She received her BA degree in Mathematics and Statistics and MBA in Marketing and HR from Andhra University, Visakhapatnam. She received a Ph.D degree in Marketing from Andhra University, Visakhapatnam, India. She has published 25 research papers in refereed international journals and 30 in various national and international conference proceedings. She also published patents. Her research areas include Supply chain Management, Human Resource Management, Consumer Behavior, Artificial Intelligence, and Business Analytics. She is an active member of the Commerce and Management Association of India (CMOAI).

Rajasekhara Mouly Potluri is a Professor of Management/Marketing with thirty-three years of industry/teaching experience in around ten countries and is currently working as a Professor of Marketing at the Business School of Kazakh British Technical University of Almaty, Kazakhstan. Dr. Raj holds a Ph.D. and M. Phil. in Management/Marketing from Shivaji University, an MBA (Marketing), and a Master of Commerce (Banking) from Andhra University, India. He has published and presented around a hundred and thirty research articles, books, book chapters, and case studies in renowned peer-reviewed journals indexed in Scopus, ABDC, SSCI, IEEE, KCI, and International Conferences and won more than twenty Best Research Paper Awards and Academic Service Excellence Awards. Prof. Raj has extensive teaching and research experience in multicultural and multiethnic environments in India, Ethiopia, Kazakhstan, South Korea, Nigeria, and the United Arab Emirates (Golden Visa Received). His research interests are Marketing, Islamic Marketing, CSR and Sustainability, HRM & OB, and Entrepreneurship. He has professional memberships in the Eurasian Business and Economics Society, International Islamic Marketing Association, Social Responsibility Research Network, and International Society for Development and Sustainability, Emerald Literate Network, and Consumer Guidance Society of India.

Problems and Prospects of AI in the Global Healthcare Sector: A Case of Kazakhstan

Rajasekhara Mouly Potluri, Yerzhan B. Mukashev, and Inzhu Baktygulova

Abstract This chapter examines the challenges and opportunities within the global healthcare sector, with a specific focus on Kazakhstan, incorporating the role of artificial intelligence (AI) in enhancing healthcare delivery. By utilizing surveys to estimate public perception and experiences within the healthcare system, this study identifies critical areas for improvement and potential strategies, including AI-driven solutions, to address key issues. Through qualitative and quantitative research methods, this chapter explores factors influencing the global healthcare sector, including technological advancements such as AI, demographic shifts, and healthcare policy reforms. This chapter provides a comprehensive analysis of Kazakhstan's healthcare landscape, integrating data from government reports, academic studies, and industry publications. Several challenges facing Kazakhstan's healthcare sector—such as inadequate infrastructure, limited access to healthcare services in rural areas, and healthcare workforce shortages—are examined in light of how AI can offer innovative solutions. For instance, AI can be leveraged to improve diagnostic accuracy, optimize resource allocation, and expand telemedicine services, especially in underserved regions. The study also highlights the transformative potential of AI in improving healthcare access and delivery, offering promising opportunities to overcome existing barriers. However, this research is limited by the availability and reliability of data, particularly concerning healthcare indicators and outcomes in Kazakhstan, as well as the complexities involved in integrating AI technologies into existing systems. By examining the problems and prospects of the global healthcare sector through the lens of Kazakhstan and incorporating AI-driven strategies, this study contributes to a deeper understanding of healthcare dynamics in both local and global contexts. The insights gained have significant implications for policymakers, healthcare professionals, and other stakeholders seeking to harness AI's potential to address challenges and capitalize on opportunities within Kazakhstan's healthcare sector.

R. M. Potluri (✉) · Y. B. Mukashev · I. Baktygulova
KBTU Business School, Kazakh-British Technical University, Almaty, Kazakhstan
e-mail: prmouly24@gmail.com

N. R. Vajjhala et al. (eds.), *Artificial Intelligence in Healthcare Information Systems—Security and Privacy Challenges*, Information Systems Engineering and Management 34, https://doi.org/10.1007/978-3-031-84404-1_10

Keywords Healthcare sector · Challenges · Opportunities · Technology · International collaborations · Kazakhstan · Artificial intelligence · Healthcare delivery · Technological advancements · Global healthcare sector

1 Introduction

The healthcare sector is a foundation of societal well-being, reflecting a critical intersection of public health, economic stability, and social equity. Across the globe, nations struggle with the evolving landscape of healthcare delivery, striving to balance the demands of rising healthcare costs, advancing medical technologies, and shifting demographics. Improving global health means ensuring that current and future medications are effective and affordable [1]. As healthcare costs rise worldwide, branded and generic drug companies are dealing with how to stand out in the market. One big issue with long-term medications is that people often don't stick to their treatment plans, and the body can sometimes develop ways to compensate for the effects of the drugs over time. This can lead to the medications becoming less effective or even ineffective altogether.

On the other hand, the global pandemic has prompted significant advancements and adaptations within the healthcare landscape, leading to overall improvements in various aspects of the healthcare sector worldwide. It includes digital healthcare, public health awareness, vaccine development, and research. The global and local dashboards that integrate data sourced from the World Health Organization (WHO) and the Centers for Disease Control and Prevention (CDC) were developed for real-time tracking of Covid-19 statistics [2]. These dashboards present comprehensive information on cases categorized by country and region, visualized through maps and charts featuring live geospatial updates for ongoing monitoring of cases. The infrastructure and technology developed for these dashboards can be repurposed and adapted to address other healthcare issues beyond the Covid-19 pandemic. For example, they can monitor the spread of other infectious diseases, such as influenza, Zika virus, or Ebola, providing real-time updates on case numbers, geographic distribution, and trends. Another application could be the monitoring and evaluation of public health interventions, such as vaccination campaigns, health education initiatives, or sanitation projects, to measure their impact on population health outcomes.

The healthcare challenges present in Kazakhstan match the global concerns outlined earlier. Like other nations, Kazakhstan struggles with high healthcare costs and accessibility issues, particularly in rural and remote regions. These challenges are compounded by disparities in healthcare access and medication adherence, reflecting broader systemic problems within the healthcare sector. Furthermore, the Covid-19 pandemic has magnified existing vulnerabilities, placing additional strain on Kazakhstan's healthcare infrastructure and increasing disparities in healthcare delivery. Despite these challenges, Kazakhstan has shown resilience and a commitment to addressing healthcare disparities through ongoing efforts to enhance infrastructure, expand access to essential services, and strengthen public health initiatives. The study of healthcare challenges and advancements, particularly within Kazakhstan, holds significant importance for several reasons. Firstly, understanding the specific healthcare issues faced by Kazakhstan provides a valuable contribution to the broader challenges faced by developing and transitional economies. Some authors have mentioned how its unique geographical characteristics challenge public health improvement within Kazakhstan. Initially, despite ranking as the 9th largest country globally and the 4th largest in Asia, Kazakhstan stands out for its notably sparse population distribution [3]. With an average density of just under 6.93 individuals per square kilometer, it ranks low at 184th in population density worldwide.

Furthermore, Kazakhstan is home to two of the world's most significant environmentally hazardous zones: the Baikonur Cosmodrome and the Semipalatinsk nuclear test site. Upon the dissolution of the USSR, ownership of the cosmodrome transferred to Kazakhstan, though it remains leased by the Russian Federation. The pollution of the biosphere resulting from hazardous rocket fuel poses a significant environmental threat. At the same time, the Semipalatinsk nuclear test site has heightened the incidence of cardiovascular diseases and cancer among residents in its location for many generations. Understanding the healthcare challenges and advancements in Kazakhstan offers significant inputs into broader issues faced by developing economies. Kazakhstan allocated 236.35 USD per person for healthcare expenditures [4]. Despite being the second highest value in Post Soviet Union countries, this amount is still very little compared to worldwide statistics. For example, China's allocation was 978.8 USD, Turkiye's was 1827.5, and the USA had the highest portion, which equals 12555.3 USD [5]. This information highlights potential challenges in adequately funding healthcare services and infrastructure in Kazakhstan, which could impact the accessibility and quality of healthcare for its population. Additionally, it shows the importance of understanding the broader context of healthcare spending and its implications for public health outcomes and healthcare system performance. Kazakhstan offers a sample of the healthcare gaps and systemic obstacles in similar regions worldwide. By examining the healthcare landscape in Kazakhstan, this study contributes to a deeper understanding of the complex interplay between socioeconomic factors, healthcare infrastructure, and public health outcomes.

2 Literature Review

A key obstacle in healthcare operations management revolves around efficiently utilizing costly resources while supporting the quality of care delivered. Simulation and optimization techniques offer effective avenues to improve healthcare services by crafting models to reduce patient wait times, minimize healthcare supply chain and logistics expenses, and optimize accessibility [6]. Companies worldwide aim to reduce environmental impact by adopting sustainable supply chain practices. In pursuit of this objective, suppliers play an essential role in implementing and maintaining a sustainable supply chain. Indian healthcare has gathered global recognition for its affordability, skilled professionals, and specialized medical services. India's healthcare expenditure as a percentage of GDP stood at approximately 3.7%, significantly lower than that of many developed countries, indicating the cost-effectiveness of its healthcare system [7]. This affordability is particularly evident in medical tourism, where India emerges as a top destination for patients seeking quality treatment at lower costs. Another study highlights India's expertise in medical tourism, attributing its success to factors such as internationally accredited hospitals, English-speaking medical professionals, and the availability of advanced medical technologies [8].

The country's medical workforce is also praised for its adaptability and resilience, as evidenced by its essential role in responding to health crises such as the COVID-19 pandemic. Moreover, another crucial study stated that India is recognized for its expertise in specialized medical fields, with institutions like the All India Institute of Medical Sciences (AIIMS) leading advancements in cardiology, oncology, and organ transplantation [9]. In her report, Lahariya mentioned that the Indian Government launched the Ayushman Bharat Program (ABP), comprising two key components, in February 2018 [10]. Firstly, the Health and Wellness Centres (HWCs) initiative aimed to provide comprehensive primary healthcare (PHC) services to the entire population. Secondly, the Pradhan Mantri Jan Arogya Yojana (PMJAY) sought to enhance access to hospitalization services at secondary and tertiary level health facilities for the bottom 40% of the total population. The launch results were stated in the same report: infrastructure strengthening and multiskilling achieved, two-way functional referral mechanisms set up, and expanded health services delivered closer to the community. Overall, Indian healthcare's affordability, skilled workforce, and specialization contribute to its reputation as a leading healthcare destination on the global stage.

Another healthcare system globally known for its efficiency, accessibility, and high-quality care belongs to Singapore. At the base of Singapore's healthcare model is universal coverage through public and private sector involvement. The Ministry of Health (MOH) oversees the healthcare system, which is funded primarily through compulsory contributions to the Central Provident Fund (CPF), a social security savings plan [11]. According to the World Bank, Singapore's healthcare expenditure as a percentage of GDP was approximately 4.7% in 2019, reflecting a balanced allocation of resources to maintain affordability and sustainability. One of the features of Singapore's healthcare system is its focus on preventive care and health promotion. The government actively promotes healthy living and disease prevention through the Health Promotion Boards (HPB) National Healthy Lifestyle Campaigns and subsidized screenings for chronic diseases like diabetes and hypertension [12]. This proactive approach has contributed to Singapore's impressive health outcomes, including one of the highest life expectancies and lowest infant mortality rates globally.

In terms of accessibility, Singapore has a comprehensive network of healthcare facilities and services that cater to its population's diverse needs. Public hospitals, polyclinics, and specialist centers provide affordable and accessible care, while private hospitals offer additional options for those seeking premium services. The combination of public and private provision ensures that Singaporeans have timely access to various medical treatments and specialists. Singapore's healthcare system is also recognized for its efficiency and innovation. The introduction of Medisave, MediShield, and MediFund schemes has helped alleviate patients' financial burdens, ensuring that healthcare remains affordable and accessible to all income groups [13]. Moreover, adopting technology and data analytics has enhanced healthcare delivery, with initiatives like the National Electronic Health Record (NEHR) system facilitating seamless information exchange between healthcare providers and improving patient outcomes. Another academician has suggested that Singapore's healthcare system stands out for its comprehensive coverage, emphasis on preventive care, accessibility, efficiency, and innovation [14]. These attributes have earned Singapore a reputation as a global leader in healthcare delivery and serve as a model for other countries seeking to achieve international excellence in healthcare.

Compared to the best healthcare providers worldwide, the healthcare system in Kazakhstan is still transforming, facing both challenges and prospects as it attempts to meet the healthcare needs of its population. Kazakhstan's healthcare system is primarily funded by the government through the National Health Insurance Fund (NHIF), supplemented by out-of-pocket patient payments [15]. However, according to the World Bank, healthcare expenditure as a percentage of GDP in Kazakhstan was relatively low at around 3.4% in 2019, indicating potential limitations in resource allocation and sustainability. One of the critical challenges facing Kazakhstan's healthcare system is the uneven distribution of healthcare resources, especially between urban and rural areas [16]. While urban centers like Nur-Sultan and Almaty boast modern healthcare facilities and specialists, rural regions often face medical personnel and infrastructure shortages. This difference in healthcare access contributes to gaps in health outcomes and health inequities among different population groups.

Another pressing issue is the quality of healthcare services in Kazakhstan. Despite efforts to improve healthcare infrastructure and training programs for healthcare professionals, concerns persist regarding the standard of care, patient safety, and medical errors. Additionally, the frequency of non-communicable diseases (NCDs) such as cardiovascular diseases, diabetes, and cancer poses significant public health challenges, requiring comprehensive prevention and management strategies [17]. However, in the middle of these challenges, Kazakhstan's healthcare system also presents opportunities for improvement and innovation. The government has implemented various reforms to strengthen primary healthcare and enhance healthcare delivery at the grassroots level. Initiatives such as the "Densaulyk" state healthcare development program prioritize preventive care, health promotion, and early detection of diseases [18].

Furthermore, Kazakhstan's strategic location at the intersection of Europe and Asia positions it as a potential medical tourism and healthcare investment sample. Some academicians argued that with its modern healthcare infrastructure, skilled workforce, and relatively low healthcare costs compared to Western countries, Kazakhstan has the potential to attract patients from neighboring countries and beyond, thereby stimulating economic growth and international collaboration in the healthcare sector [19]. In conclusion, Kazakhstan's healthcare system faces challenges related to resource allocation, healthcare access, and quality of care. However, there are also promising opportunities for reform and innovation, including strengthening primary healthcare, addressing health disparities, and leveraging medical tourism potential.

2.1 Theoretical/Conceptual Framework

Understanding the problems and prospects of the global healthcare sector, particularly within Kazakhstan, can be improved by integrating insights from established theories and frameworks: Whitehead's Health Equity Framework emphasizes the importance of addressing social determinants of health and reducing health inequities within populations [20]. This framework highlights the distribution of resources, opportunities, and power as crucial determinants of health outcomes. In the context of Kazakhstan, applying this framework involves identifying and addressing inequities in healthcare access, quality, and outcomes among different population groups, including rural versus urban areas, ethnic minorities, and low-income communities, and also implementing policies that promote equitable distribution of healthcare resources, such as healthcare facilities, healthcare professionals, and essential medicines, to underserved regions and populations—addressing underlying social determinants of health, including poverty, education, employment, housing, and environmental factors, through multisectoral collaboration and community engagement initiatives [21].

Resource Dependency Theory examines the relationships between organizations and their external environments, focusing on acquiring and managing resources necessary for organizational survival and success [22]. In the healthcare sector of Kazakhstan, this theory can be applied to analyze the interdependencies between healthcare providers, payers, regulators, and other stakeholders, recognizing that healthcare organizations rely on external resources such as funding, technology, and expertise to deliver quality care. Another application is assessing the vulnerabilities and dependencies of the healthcare system on external factors such as government policies, economic conditions, and global health trends and developing strategies to mitigate risks and enhance resilience [23].

Finally, Donabedian's Quality Improvement Framework focuses on assessing and improving healthcare quality by evaluating structure, process, and outcome measures [24]. In Kazakhstan's healthcare sector, this framework assesses the structure of healthcare delivery systems, including the availability of healthcare facilities, equipment, and personnel, to ensure adequate capacity and readiness to meet population health needs. Monitoring and improving healthcare processes, such as clinical guidelines adherence, patient safety protocols, and care coordination mechanisms, to enhance the efficiency and effectiveness of care delivery [25]. Evaluating healthcare outcomes, including patient health status, satisfaction levels, and health disparities, to measure the impact of quality improvement efforts and guide continuous performance improvement is imperative to any country's healthcare sector.

By integrating the Health Equity Framework, Resource Dependency Theory, and Quality Improvement Framework into the analysis, Kazakhstan stakeholders can better understand the challenges and opportunities facing the healthcare sector and design targeted strategies to promote health equity, optimize resource allocation, and enhance healthcare quality. This theoretical/conceptual framework serves as a guiding framework for research, policy development, and action aimed at advancing the goals of universal health coverage, health equity, and sustainable development in Kazakhstan and beyond.

2.2 *Hypotheses*

H1: There is a significant association between socioeconomic status and access to healthcare services in Kazakhstan, with individuals from lower socioeconomic backgrounds experiencing more significant barriers to healthcare access than those from higher socioeconomic backgrounds.

Drawing from the Health Equity Framework, which emphasizes the importance of addressing social determinants of health, this hypothesis states that socioeconomic disparities contribute to differential access to healthcare services in Kazakhstan. Lower-income individuals may face financial, geographic, and cultural barriers to accessing healthcare, leading to inequities in healthcare utilization and health outcomes. By testing this hypothesis, policymakers can identify priority areas for intervention and develop targeted strategies to improve healthcare access and reduce population gaps.

H2: Healthcare organizations in Kazakhstan with diversified funding sources and strategic partnerships with external stakeholders exhibit greater resilience and adaptability in response to external shocks, such as economic downturns or changes in government policies, compared to organizations that rely heavily on a single funding source or stakeholder.

Resource Dependency Theory suggests that organizations depend on external resources for survival and success. In the healthcare context in Kazakhstan, this hypothesis posits that organizations with diversified funding sources and strategic partnerships are better equipped to navigate uncertainties and mitigate risks associated with external dependencies. By testing this hypothesis, policymakers and healthcare managers can assess the management of healthcare organizations and develop strategies to improve resource diversification and collaboration to promote sustainability and continuity of care delivery.

H3: Implementation of evidence-based quality improvement initiatives in healthcare facilities in Kazakhstan is associated with improvements in patient outcomes, including reduced hospital-acquired infections, decreased medication errors, and increased patient satisfaction levels.

The Quality Improvement Framework emphasizes the importance of systematic approaches to assessing and improving healthcare quality. This hypothesis suggests that healthcare facilities in Kazakhstan that prioritize quality improvement efforts, such as implementing clinical guidelines, patient safety protocols, and continuous performance monitoring, will experience positive impacts on patient outcomes. By testing this hypothesis, healthcare administrators and quality improvement professionals can evaluate the effectiveness of quality improvement initiatives and identify areas for further enhancement to optimize patient care and safety.

3 Research Methodology

The research study was conducted using a quantitative approach, specifically through the conduction of surveys. Surveys are efficient and cost-effective for gathering large-scale data from a diverse population. The survey instrument was designed to collect structured data on critical variables related to the problems and prospects of the healthcare sector in Kazakhstan, as outlined within the Health Equity Framework, Resource Dependency Theory, and Quality Improvement Framework. The survey captured information on various aspects of healthcare access, quality, and equity, as well as organizational and external factors influencing healthcare delivery and outcomes. Questions were formulated to assess respondents' perceptions, experiences, and opinions regarding healthcare access, resource allocation, quality improvement efforts, and health outcomes in Kazakhstan. The survey instrument was administered after testing its validity and reliability to ensure the reliability and validity of the data collected. Using surveys as the primary data collection method, the research study will provide an essential quantitative contribution to the challenges and opportunities facing the healthcare sector in Kazakhstan. The primary data for this research study was collected directly from respondents through surveys administered via Google Forms. This questionnaire is meant to gather firsthand information from respondents regarding their perceptions, experiences, and opinions related to the healthcare sector in Kazakhstan. The secondary data collected chiefly from the data sources, such as existing literature, reports, and statistical databases, will also be utilized to provide additional context and background information on the healthcare sector in Kazakhstan. The research approach employed in this study was primarily survey research, using structured questionnaires administered to a sample of respondents. This approach allows for systematic data collection from a large and diverse population, providing quantitative insights into the research topic. The primary research instrument for data collection was online questionnaires in English and Russian and distributed through Google Forms. These questionnaires were designed to capture quantitative data on various aspects of the healthcare sector in Kazakhstan, as outlined in the research objectives. Additionally, qualitative measures may be incorporated to gather more nuanced insights into respondents' perceptions and experiences.

This study's sampling unit represents diverse stakeholder groups within the healthcare sector in Kazakhstan, including healthcare providers, administrators, policymakers, and patients. The sample size was 300 respondents targeted for the survey. This sample size is sufficient to achieve a reasonable statistical power and precision level while ensuring practical feasibility and resource constraints, and the sampling procedure respondents selected through convenience sampling, whereby individuals with access to the survey link will be invited to participate. Recruitment efforts have been made through various channels, including professional networks, social media platforms, and healthcare organizations. The targeted respondents were primarily contacted through online channels, specifically email and social media platforms. The survey link was distributed electronically to reach a broad audience efficiently and cost-effectively, and respondents were contacted through personal networks and professional associations to enhance participation rates and sample representativeness. The scope of this research study encompasses examining the problems and prospects of the healthcare sector in Kazakhstan, focusing on understanding critical issues related to health equity, resource dependency, and quality improvement. Specifically, the study aims to investigate inequities in healthcare access, quality, and outcomes among Kazakhstan's population, particularly vulnerable and marginalized populations. Analysis of the factors contributing to resource dependencies within the healthcare sector, including financial, technological, and organizational dependencies, and assess their implications for healthcare delivery and sustainability. The research study gathers data from diverse stakeholders, including healthcare providers, administrators, policymakers, and patients, to comprehensively understand Kazakhstan's healthcare landscape. Survey data was analyzed using quantitative and qualitative techniques to derive meaningful insights and conclusions. The methods such as descriptive statistics, inferential statistics, and content analysis will be employed. Descriptive statistics, such as frequencies, percentages, means, and standard deviations, are administered to summarize and describe the characteristics of the study sample and critical variables related to healthcare access, resource dependency, and quality improvement. Inferential statistics, including correlation analysis, regression analysis, and hypothesis testing, examined relationships and associations between variables of interest. These statistical tests helped to identify significant data predictors, trends, and patterns. Content analysis provided additional context and background information to support the quantitative findings. Lastly, the quantitative and qualitative findings are integrated through a triangulation approach, whereby quantitative results are compared with qualitative insights to provide a deeper understanding.

4 Results and Discussion

4.1 Demographic Profile of the Subjects

Table 1 provides the demographic profile of the subjects, in the context of the age. The data provides a distribution of respondents across different age groups, indicating the age diversity within the sample population. The age group "26–35" has the highest number of respondents (92), suggesting that individuals in this age bracket are more likely to participate in the survey.

There is a relatively higher representation of younger age groups (e.g., "18–25" and "26–35") compared to older age groups (e.g., "56–65" and "65 or more"), which may impact the generalizability of the findings, especially for issues that disproportionately affect older populations. Table 2 presents the demographic gender profile for the survey respondents, while Table 3 present the educational levels of the respondents.

The data provides insights into the educational attainment of the survey respondents, reflecting the diversity of academic backgrounds within the sample population. Most respondents have attained higher education qualifications, with bachelor's degrees being the most common (209 respondents), followed by master's degrees (42 respondents) and Ph.D.s (16 respondents). The higher prevalence of bachelor's, master's, and Ph.D. qualifications among respondents suggests a highly educated sample population, which may influence the interpretation of survey results and

Table 1 Demographic profile: age

Question	Response	Count of age
What is your age?	Less than 18	12
	18–25	73
	26–35	92
	36–45	50
	46–55	33
	56–65	29
	65 or more	11
Total		300

Source Survey results

Table 2 Demographic profile: gender

Question	Response	Count of gender
Gender	Female	163
	Male	137
Total		300

Source Survey results

Table 3 Demographic profile: education level

Question	Response	Count of education
What is your highest education level?	Elementary school	0
	Middle school	3
	High school	12
	Bachelor's degree	209
	Master's degree	42
	Ph.D.	16
	Other	18
Total		300

Source Survey results

Table 4 Reliability of the questionnaire

Question	Response	Count of rating
How would you rate the relevance of the survey questions in capturing your experiences and opinions regarding the healthcare sector in Kazakhstan?	1	13
	2	24
	3	69
	4	135
	5	59
Total		300

Source Survey results

the analysis of healthcare-related perspectives and experiences. Table 4 present the reliability results of the survey results.

While a substantial number of respondents rated the survey questions as either somewhat relevant (135 respondents) or highly relevant (59 respondents), some respondents rated the questions as somewhat irrelevant (24 respondents) or not appropriate at all (13 respondents). The most common response was a neutral rating (69 respondents), indicating that a significant portion of respondents neither strongly agreed nor disagreed with the relevance of the survey questions.

4.2 *Testing of Hypotheses*

Table 5 presents the hypothesis testing results for H1.

Many respondents (108 + 103 = 211) either agree or strongly agree with disparities in healthcare access among different population groups in Kazakhstan, indicating a widespread perception of unequal access to healthcare services. Table 6 presents the hypothesis testing results for H2.

Table 5 Testing of Hypothesis 1

Question	Response	Count of rating
To what extent do you perceive disparities in healthcare access among different population groups in Kazakhstan?	1	21
	2	26
	3	42
	4	108
	5	103
Total		300

Source Survey results

Table 6 Testing of Hypothesis 2

Question	Response	Count of rating
How would you rate the availability of healthcare resources (e.g., facilities, equipment, personnel) in your region of Kazakhstan?	1	53
	2	42
	3	143
	4	38
	5	24
Total		300

Source Survey results

The most common perception among respondents is moderate availability (143 out of 300), indicating that a significant portion of respondents view healthcare resource availability in their region as neither exceptionally low nor high. While a considerable number of respondents perceive the availability of healthcare resources as low or deficient ($53 + 42 = 95$), some respondents perceive it as high or very high ($38 + 24 = 62$), suggesting both areas of concern. Table 7 presents the results of the hypothesis for H3.

The popular view among respondents is moderate effectiveness (124 out of 300), indicating that a significant portion of participants consider quality improvement efforts to have a mild impact on addressing healthcare challenges and improving patient care. However, a noteworthy number of respondents perceive these efforts as ineffective or highly ineffective ($58 + 69 = 127$), suggesting significant concerns regarding the effectiveness of current quality improvement initiatives in Kazakhstan. Table 8 presents the regression analysis results of the hypotheses.

H1: The statistical significance ($P < 0.05$) suggests a significant association between socioeconomic status and access to healthcare services in Kazakhstan. This result implies that individuals from lower socioeconomic backgrounds are more

Table 7 Testing of Hypothesis 3

Question	Response	Count of rating
How would you assess the effectiveness of quality improvement efforts in addressing healthcare challenges and enhancing patient care in Kazakhstan?	1	58
	2	69
	3	124
	4	36
	5	13
Total		300

Source Survey results

Table 8 Regression analysis of hypotheses

Hypothesis	Methodology	Result	Conclusion
H1	Regression analysis	$P < 0.05$, $R^2 = 0.26$	Substantial evidence supports the hypothesis that there is a relationship between the variables being studied. The independent variable(s) included in the model explain around 26% of the variability in the dependent variable. This result suggests a moderate level of explanatory power in the model
H2	Regression analysis	$P < 0.05$, $R^2 = 0.27$	There is a statistically significant relationship between the variables being studied. The regression model's independent variable (s) can explain approximately 27% of the variability in the dependent variable
H3	Regression analysis	$P < 0.05$, $R^2 = 0.48$	There is a statistically significant relationship between the variables being studied. The regression model's independent variable (s) can explain approximately 48% of the variability in the dependent variable

Source Survey results and regression analysis

likely to face barriers to healthcare services than those from higher socioeconomic backgrounds. The moderate explanatory power of the model ($R^2 = 0.27$) indicates that socioeconomic status explains approximately 27% of the variability in access to healthcare services.

H2: The statistical significance ($P < 0.05$) suggests a significant relationship between funding diversification and strategic partnerships with external stakeholders and organizational resilience and adaptability in Kazakhstan's healthcare organizations. This theory implies that organizations with diversified funding sources and strategic alliances are more resilient and adaptable to external shocks than those relying heavily on a single source. The moderate explanatory power of the model ($R^2 = 0.36$) indicates that funding diversification and strategic partnerships explain approximately 36% of the variability in organizational resilience and adaptability.

Table 9 Findings

Findings	Discussion
Overall age distribution	The distribution of respondents across different age groups provides insights into the demographic profile of the sample population. It may influence the interpretation of survey results, particularly regarding age-related perspectives and experiences related to the healthcare sector
Educational profile	While bachelor's, master's, and Ph.D. qualifications are predominant, a small proportion of respondents (33) reported having completed elementary, middle, or high school education, indicating a range of educational backgrounds represented in the sample population
Impact on survey quality	Understanding respondents' perceptions of the relevance of survey questions is essential for assessing the overall quality of the survey instrument and ensuring that it effectively captures the intended information and perspectives related to the healthcare sector in Kazakhstan
Room for improvement	While some respondents perceive quality improvement efforts as practical or very effective ($36 + 13 = 49$), there is still room for improvement, as indicated by the substantial number of respondents who perceive these efforts as moderately effective or ineffective

Source Survey results

H3: The statistical significance ($P < 0.05$) suggests a significant association between implementing evidence-based quality improvement initiatives and improving patient outcomes in Kazakhstan's healthcare facilities. This hypothesis implies that facilities that implement such initiatives experience positive changes in patient outcomes. The relatively high explanatory power of the model ($R^2 = 0.48$) indicates that the implementation of evidence-based quality improvement initiatives explains approximately 48% of the variability in patient outcomes.

4.3 Discussion and Presentation of Findings

Table 9 preents a summary of the findings of this study.

5 Managerial Implications

Integrating the stakeholder-wise analysis with the results of the previous analysis provides a comprehensive understanding of the managerial implications for healthcare leaders in Kazakhstan. By considering both the statistical findings and the needs of various stakeholders, managers can develop strategies to address healthcare challenges effectively.

Government and Policy Makers: Building upon the statistical evidence of disparities in healthcare access (H1), government and policymakers can prioritize initiatives to reduce inequalities identified through stakeholder-wise analysis. This tendency may involve targeted funding allocations for healthcare infrastructure development in underserved regions and implementing policies that encourage healthcare providers to address the needs of vulnerable populations.

Healthcare Organizations: Leveraging the findings of organizational resilience and adaptability (H2), healthcare organizations can align their strategic priorities with stakeholder expectations. Managers can use stakeholder-wise analysis to identify key funding partners and collaborative opportunities that support their resilience efforts.

Healthcare Providers: Integrating evidence-based quality improvement initiatives (H3) with stakeholder-wise analysis allows healthcare providers to tailor their approaches to meet the diverse needs of patients and communities. Managers can prioritize quality improvement efforts based on stakeholder feedback, focusing on areas identified as critical for improving patient outcomes.

Patients and Communities: Recognizing the impact of socioeconomic factors on healthcare access and outcomes, managers can develop targeted interventions informed by stakeholder-wise analysis. By integrating community outreach programs and health education initiatives with statistical evidence of disparities in healthcare access (H1), managers can address the root causes of health inequities and empower patients to take control of their health.

6 Conclusion, Limitations, and Scope for Further Research

Overall, this study has provided significant input into the challenges and prospects of the healthcare sector in Kazakhstan from a global perspective. Through a rigorous analysis of key factors such as socioeconomic status, funding diversification, and quality improvement initiatives, significant findings have emerged that can inform strategic decision-making and policy formulation. Based on the results, it is recommended that stakeholders in the healthcare ecosystem, including government agencies, healthcare organizations, providers, and communities, collaborate to address the identified challenges and capitalize on the opportunities highlighted in this study. Initiatives to reduce healthcare access disparities, diversify funding sources, and implement evidence-based quality improvement measures should be prioritized to enhance patient outcomes.

Despite the gained inputs above, it is essential to acknowledge the limitations of this study. One limitation is the reliance on survey data, which may be subject to response bias and limited generalizability. Additionally, the study focused primarily on quantitative analysis, which may not capture the full complexity of healthcare challenges and prospects in Kazakhstan. Qualitative research methods like interviews and focus groups could provide deeper insights into stakeholders' perspectives and experiences. Furthermore, the study's scope was limited to specific variables and hypotheses, neglecting other potentially relevant factors that could impact the healthcare sector in Kazakhstan. For instance, cultural and institutional factors, healthcare infrastructure, and the role of technology in healthcare delivery were not thoroughly explored. Future research should consider a broader range of variables to provide a more comprehensive understanding of the healthcare landscape in Kazakhstan.

There are several aspects for further research to build upon the findings of this study and address its limitations. Firstly, qualitative research methods can complement quantitative analysis and provide a richer understanding of stakeholders' perspectives and experiences. Additionally, understanding how cultural norms, traditions, and institutional structures influence healthcare access and utilization could inform the development of culturally competent care practices and policies tailored to the needs of diverse populations. Furthermore, longitudinal studies tracking changes in healthcare indicators over time could improve trends and patterns in the healthcare sector. By monitoring key metrics such as healthcare access, quality of care, and patient outcomes, researchers can assess the effectiveness of interventions and policy initiatives to improve Kazakhstan's healthcare system.

In conclusion, while this study has highlighted important issues within the healthcare sector in Kazakhstan, much remains to be explored. By addressing its limitations and embracing opportunities for further research, scholars and practitioners can contribute to ongoing efforts to enhance healthcare delivery and outcomes in Kazakhstan and beyond.

References

1. Ilan, Y.: Improving global healthcare and reducing costs using second-generation artificial intelligence-based digital pills: a market disruptor. Int. J. Environ. Res. Public Health **18**(2), 811 (2021)
2. Kapoor, A., Guha, S., Das, M.K., Goswami, K.C., Yadav, R.: Digital healthcare: The only solution for better healthcare during the COVID-19 pandemic? Indian Heart J. **72**(2), 61–64 (2020)
3. Gulis, G., Aringazina, A., Sangilbayeva, Z., Zhan, K., de Leeuw, E., Allegrante, J.P.: Population health status of the republic of Kazakhstan: trends and implications for public health policy. Int. J. Environ. Res. Public Health **18**(22), 12235 (2021)
4. Semenova, Y., Lim, L., Salpynov, Z., Gaipov, A., Jakovljevic, M.: Historical evolution of healthcare systems of Post-Soviet Russia, Belarus, Kazakhstan, Kyrgyzstan, Tajikistan, Turkmenistan, Uzbekistan, Armenia, and Azerbaijan: a scoping review. Heliyon (2024)
5. Statista.: J. Bus. & Finance Librarian. **27**(4), 304–309 (2022)
6. Bhosekar, A.S.: Essays on perioperative services problems in healthcare (2022)
7. The World Bank, World Development Indicators: GDP per capita, Atlas method. [Data file] (2019). Retrieved from http://data.worldbank.org/indicator
8. Biswas, S.: Measuring the performance of healthcare supply chains in India: a comparative analysis of multi-criteria decision-making methods. Decis. Mak.: Appl. Manag. Eng. **3**(2), 162–189 (2020)
9. Suryavanshi, N., Kadam, A., Dhumal, G., Nimkar, S., Mave, V., Gupta, A., Gupte, N.: Mental health and quality of life among healthcare professionals during the COVID-19 pandemic in India. Brain Behav.Behav. **10**(11), e01837 (2020)
10. Lahariya, C.: Health & wellness centers to strengthen primary health care in India: concept, progress, and ways forward. Ind. J. Pediatrics **87**(11), 916–929 (2020)
11. Teo, I., Chay, J., Cheung, Y.B., Sung, S.C., Tewani, K.G., Yeo, L.F., Yang, G.M., Pan, F.T., Ng, J.Y., Abu Bakar Aloweni, F., Ang, H.G., Tan, H.K.: Healthcare worker stress, anxiety

and burnout during the COVID-19 pandemic in Singapore: a 6-month multi-center prospective study. PLoS ONE **16**(10), e0258866 (2021)
12. Emanuel, E.J.: Which Country Has the world's Best Health Care? Hachette UK (2020)
13. Nagar, S.: The struggle of mental health care delivery in South Korea and Singapore. Harv. Int. Rev.. Int. Rev. **43**(1), 39–43 (2022)
14. Muneeswaran, V., Nagaraj, P., Dhannushree, U., Ishwarya Lakshmi, S., Aishwarya, R., Sunethra, B.: A framework for data analytics-based healthcare systems. In: Innovative Data Communication Technologies and Application: Proceedings of ICIDCA 2020, pp. 83–96. Springer (2021)
15. Koichubekov, B., Kharin, A., Sorokina, M., Korshukov, I., Omarkulov, B.: System dynamics modeling for general practitioner workforce forecasting in Kazakhstan. Ann. Ig. **33**(3), 242–253 (2021)
16. Jobalayeva, B., Khismetova, Z., Glushkova, N., Kozhekenova, Z., Abzaliyeva, A., Berikuly, D., Semenova, Y.: The impact of incentive scheme on rural healthcare workforce availability: a case study of Kazakhstan. Hum. Resour. HealthResour. Health **22**(1), 1–11 (2024)
17. Shaltynov, A., Rocha, J., Jamedinova, U., Myssayev, A.: Assessment of primary healthcare accessibility and inequality in north-eastern Kazakhstan. Geospatial Health **17**(1) (2022)
18. Seleznev, I., Alibekova, R., Clementi, A.: Patient satisfaction in Kazakhstan: looking through the prism of patient healthcare experience. Patient Educ. Couns.Couns. **103**(11), 2368–2372 (2020)
19. Aman, R., Ahokangas, P., Zhang, X.: Migrant women entrepreneurs and entrepreneurial ecosystems during an external shock: a case study from the healthcare sector in Kazakhstan. Asian Bus. Manag.Manag. **20**, 518–548 (2021)
20. Diderichsen, F., Whitehead, M., Dahlgren, G.: Planning for health equity in the crossfire between science and policy. Scand. J. Public Health **50**(7), 875–881 (2022)
21. Dahlgren, G., Whitehead, M.: The Dahlgren-Whitehead model of health determinants: 30 years on and still chasing rainbows. Public Health **199**, 20–24 (2021)
22. Kvangraven, I.H.: Beyond the stereotype: restating the relevance of the dependency research program. Dev. Chang. **52**(1), 76–112 (2021)
23. Barney, J.B., Ketchen, D.J., Jr., Wright, M.: Resource-based theory and the value creation framework. J. Manag.Manag. **47**(7), 1936–1955 (2021)
24. White, K.M., Dulko, D., DiPietro, B.: The effect of burnout on quality of care using Donabedian's framework. Nurs. Clin.. Clin. **57**(1), 115–130 (2022)
25. Edwards, K.H., FitzGerald, G., Franklin, R.C., Edwards, M.T.: Air ambulance outcome measures using Institutes of Medicine and Donabedian quality frameworks: protocol for a systematic scoping review. Syst. Rev. **9**, 1–8 (2020)

Rajasekhara Mouly Potluri is a Professor of Management/Marketing with thirty-three years of industry/teaching experience in around ten countries and is currently working as a Professor of Marketing at the Business School of Kazakh British Technical University of Almaty, Kazakhstan. Dr. Raj holds a Ph.D. and M. Phil. in Management/Marketing from Shivaji University, an MBA (Marketing), and a Master of Commerce (Banking) from Andhra University, India. He has published and presented around a hundred and thirty research articles, books, book chapters, and case studies in renowned peer-reviewed journals indexed in Scopus, ABDC, SSCI, IEEE, KCI, and International Conferences and won more than twenty Best Research Paper Awards and Academic Service Excellence Awards. Prof. Raj has extensive teaching and research experience in multicultural and multiethnic environments in India, Ethiopia, Kazakhstan, South Korea, Nigeria, and the United Arab Emirates (Golden Visa Received). His research interests are Marketing, Islamic Marketing, CSR and Sustainability, HRM & OB, and Entrepreneurship. He has professional memberships in the Eurasian Business & Economics Society, International Islamic Marketing Association, Social Responsibility Research Network, International Society for

Development & Sustainability, Emerald Literate Network, and Consumer Guidance Society of India.

Dr. Yerzhan B. Mukashev Dr. Yerzhan B. Mukashev is a Dean and Associate Professor of the School of Business at the Kazakh-British Technical University of Kazakhstan and holds a Ph.D. in Economics from the University of Maryland, USA. He is a voracious researcher with more than two decades of experience and has published many research papers, book chapters, and international conference presentations in his area of specialization. His global exposure and extensive experience in multicultural and multi-ethnic work environments greatly benefit the organizations he works for in achieving global collaborations and flawless administration. Dr. Yerzhan's research interests are comparative institutional economics, macroeconomics, project management, entrepreneurship, and small business management.

Inzhu Baktygulova is a voracious business courses researcher at the Business School of the Kazakh British Technical University of Kazakhstan. She is an extensive researcher in disciplines like management, innovation management, artificial intelligence (AI), and international business. She is working on two projects focusing on the healthcare and education sectors, which are concerned with human resources, innovation management, and other management disciplines.

Transforming Healthcare: The Role of Health Informatics and Provider Perspectives

Philip Eappen, Ian Stedman, and Virginia Gunn

Abstract This chapter explores the diverse perspectives of healthcare providers on health informatics, highlighting both the potential benefits and the persistent challenges associated with its implementation. While health informatics offers significant advantages such as enhanced patient care, improved communication, and streamlined workflows, issues related to data security and health provider acceptance remain as critical barriers. Through a comprehensive review of the literature, the chapter identifies three key areas that must be addressed to facilitate effective adoption: training, infrastructure, and institutional policies. By emphasizing the importance of tailored training programs to boost user confidence, robust technological infrastructure to ensure data safety and integration, and clear institutional guidelines to promote accountability, this chapter advocates for a holistic approach to health informatics. Ultimately, addressing these challenges can facilitate the creation of an environment conducive to the successful integration of health informatics into clinical practice, supporting improved healthcare delivery and patient outcomes.

Keywords Health informatics · Provider perspectives · Patient care · Technology adoption · Healthcare delivery · Challenges and barriers · User engagement

1 Introduction

Information systems have been widely acknowledged for their critical role in improving and simplifying healthcare delivery [1, 2]. The incorporation of information technology into healthcare systems has driven significant advancements in global healthcare accessibility and helped resolve numerous challenges in the healthcare industry [3, 4]. Health informatics refers to the interdisciplinary field focused on

P. Eappen (✉) · V. Gunn
Cape Breton University, Sydney, NS, Canada
e-mail: philip_eappen@cbu.ca

I. Stedman
York University, Toronto, ON, Canada

N. R. Vajjhala et al. (eds.), *Artificial Intelligence in Healthcare Information Systems—Security and Privacy Challenges*, Information Systems Engineering and Management 34, https://doi.org/10.1007/978-3-031-84404-1_11

the effective use of bioscientific data and knowledge to support scientific research, strategic decision-making, and problem-solving to improve healthcare [5]. It also involves utilizing data and technology to enhance public health, healthcare systems, and patient outcomes.

The integration of health informatics with analytics is recognized for its potential to deliver personalized healthcare solutions around the world [6, 7]. Health informatics promotes collaborative care by bringing together various healthcare areas, such as ambulatory, surgery, labs, radiology, administration, and accounting [8]. Additionally, the role of health informatics in providing high-quality, efficient care has been realized through partnerships among healthcare providers, researchers, and stakeholders who embrace the principles of a continuously learning healthcare system. These technologies help both healthcare providers and patients with obtaining accurate diagnoses and support in treatment decisions [9].

Technological advancements in healthcare informatics have rapidly accelerated the development of healthcare systems globally and the use of big data in healthcare has made it easier for healthcare professionals to perform their tasks and improve care delivery [10]. Big data is applied in electronic health records, personal health records, and data from health tools like smart apps and wearable health monitors [11]. These advanced technologies are becoming more and more common in hospitals, and healthcare systems are investing heavily to expand their use to enhance healthcare practice efficiency and improve patient outcomes.

Furthermore, innovations such as integrating robotics into healthcare have advanced rapidly, with robots taking on more complex tasks, including surgeries and medical procedures. AI-powered machines, designed to replicate human cognitive functions, are also progressing [1]. Many healthcare systems have adopted AI to try to address long-standing challenges in the industry, but it is important to consider the perspectives of healthcare professionals as these technologies are being leveraged. This literature review seeks to explore healthcare providers' views on the adoption, usage, and impact of health informatics and its applications.

2 Methodology

This chapter is based on a narrative review of literature conducted between July and September 2024. We performed unstructured searches across common literature databases such as PubMed, PsycINFO, CINAHL, Scopus, and Google Scholar without a predefined protocol. While recent publications were prioritized due to the relevance of the subject, no filters were applied to limit the publication dates. The review deliberately included sources from a range of health-related disciplines and prioritized review articles over individual studies to ensure a broader, synthesized perspective rather than narrow, specific viewpoints. The search strategy comprised of keywords such as Health Informatics, Provider Perspectives, Patient Care, Technology Adoption, Healthcare Delivery, Challenges and Barriers, and synonyms.

Multiple types of studies reviewed are systematic reviews, scoping reviews, qualitative studies, and survey analyses. The settings of interest in the sources reviewed were hospitals, primary care, acute tertiary care, outpatient settings, tertiary academic institutions, veteran affairs medical center and online platforms. The included sources are from the United States (US), United Arab Emirates (UAE), and Canada. The search was not meant to be exhaustive but to provide an overview of common provider perspectives.

3 Background

3.1 Integration and Utilization of Health Informatics

The integration of health information technologies into healthcare systems provides a transformative approach aimed at enhancing patient care, streamlining workflows, and enabling more data-driven decision-making processes [4]. Several concerns associated with this integration for healthcare practitioners were raised, including increased workload and inefficiencies [11]. In particular, they identified that healthcare professionals often feel burdened by the additional tasks required to manage new technological systems. The study emphasized the importance of regulatory frameworks and incentives for vendors of health informatics tools to improve the accessibility and usability of their systems. Such proactive measures could play a crucial role in facilitating the adoption of innovations like electronic health records, making the integration process smoother for healthcare systems. The lack of training opportunities is also mentioned by the authors as contributing to providers' feeling overwhelmed by the introduction of new technologies. A range of concerns regarding training challenges (e.g., fragmented training, lack of contextual training) making it difficult for health workers to adopt health informatics technologies are also recognized in a recent review of health informatics literature [12].

Researchers identified multiple challenges in adopting health informatics, including disruptions to existing workflows, insufficient knowledge among staff, communication breakdowns, and increased operational costs [13]. Several concerns were also highlighted around data confidentiality and security, which could be significant barriers to widespread adoption [13]. Similarly, other studies explained that the lack of proper training hinders institutional flexibility and creates staff resistance to adopting new health informatics technologies. Staff resistance in adapting new innovations in healthcare was linked to insufficient preparation or education regarding the use of these systems, making it difficult for healthcare staff to effectively integrate them into their daily clinical operations. High workloads and frequent workflow interruptions by colleagues are major obstacles in integrating healthcare informatics technologies into clinical practice [14]. Interruptions such as workflow challenges can slow down processes and reduce the effectiveness of the new informatics systems. Despite these challenges, many healthcare providers have positive views regarding the role of health informatics in improving patient care and increasing clinical productivity. These positive perceptions are seen as key drivers in pushing forward the adoption of health informatics technologies.

According to Gagnon et al. [14] healthcare informatics technologies have been shown to enhance patient self-awareness, enabling individuals to take a more active role in managing their health. The authors identified system compatibility, inter-system functionality, and technological proficiency as essential factors that facilitate the successful integration of health informatics. In addition to these technical factors, organizational characteristics such as access to resources, materials, and a skilled workforce were highlighted as important determinants of successful technology adoption. Mackert et al. [15] emphasized the need to explore the connection between health literacy and the adoption of health information tools. Specifically, the authors argue that healthcare systems must ensure that both patients and healthcare providers have the necessary knowledge to use these tools effectively to fully realize the benefits of the technology. Additionally, maintaining privacy and security is critical to protect sensitive health information, and privacy and security aspects must be integrated into the adoption strategy.

Healthcare organizations are attractive targets for cyberattacks by bad actors. Major ransomware attacks and data breaches have become much too common, due in part to the belief that healthcare organizations can make financially lucrative targets [16]. Clinicians working in digitized health environments are certainly aware of these risks and that cyber breaches can have a negative impact on both patient care and patient trust. According to a study by [17], clinicians feel there is real urgency to implementing cybersecurity measures in healthcare despite the risk of short-term resource constraints. Mackert et al. [15] also recommended that sufficient training be provided to healthcare staff to better support the integration of health informatics, ensuring that these technologies are used to their fullest potential without compromising care quality or efficiency [15]. This sentiment is mirrored with respect to cybersecurity, but with the caveat that cyber fatigue is also possible. When cyber systems are properly maintained and do work well, regular updates to operational requirements will bring regular training needs. Healthcare workers want to comply but can also be confronted with legacy systems that are under-supported, thus increasing the overall burden on those tasked with monitoring risk [17].

4 Discussion

4.1 Perspectives on Big Data

Big data is increasingly recognized for its capacity to significantly improve both the quality and accessibility of patient data, providing enhanced insights and aiding in clinical decision-making. Krousel-Wood et al. [18] emphasized the importance of understanding healthcare professionals' perceptions regarding the potential benefits of big data. The study suggested that gaining a more detailed understanding of these perceptions could offer valuable insights into the ways in which big data technologies are applied within healthcare settings, and it will inform the development of more effective strategies for integrating big data into daily healthcare practices. Understanding perceptions will bridge the gap between technology implementation and its practical use and ensure that the full potential of big data is realized in improving patient care and operational efficiency.

In a case where providers transitioned from a regional electronic health record (EHR) system to a larger, commercial EHR platform, Krousel-Wood et al. [18] observed changes in healthcare professionals' perceptions of EHR benefits. The study noted areas where healthcare professionals who held favorable views of the advantages of EHRs increased, highlighting growing support for the system's potential to enhance care. However, the study also revealed areas where the proportion of supportive providers declined, indicating that some challenges persisted post-implementation, such as increased complexity or changes in workflow that were not universally well-received. Rosenkrantz et al. [19] examined the perspectives of radiologists and emergency medicine (EM) providers regarding the value of external image (diagnostic tests performed outside a given health setting) sharing in patient care. Both groups strongly perceived that external image sharing had substantial value for improving patient outcomes. The authors emphasized that the integration of outside imaging into the regional EHR system would streamline workflows, reduce redundancy, and save time, particularly for radiologists and other healthcare professionals. By making external imaging data more easily accessible, the study argued that this integration would eliminate unnecessary steps in the workflow and avoid duplication of effort, thereby improving both patient care and healthcare efficiency.

Furthermore, Kirkendall et al. [13] assessed healthcare professionals' perceptions, experiences, and expectations during the transition from a computerized provider order entry system to an EHR system. Kirkendall et al. [13] revealed that, despite initial challenges, most healthcare providers held positive views about the switch to EHRs. Many respondents felt that EHRs offered better functionality, and a much more streamlined workflow compared to the previous systems, even though the transition period involved a learning curve and some disruptions. In addition, Dash et al. [11] focused on the ongoing perceptions of care providers following the implementation of the Meaningful Use EHR (MU-EHR) system. This study found that while providers experienced a sustained increase in workload, they also noted improvements in ease of use and operational efficiency over time. Dual experience of both elevated workloads and smoother operations suggests that EHR systems might initially add complexity; however, they can eventually contribute to more efficient care delivery once healthcare providers become more accustomed to the new workflows. Dash et al. [11] findings underscore the importance of addressing workload concerns during the integration of new technologies to ensure long-term acceptance and successful use of health informatics systems.

4.2 Telehealth Innovations and Intelligence Healthcare Technologies

Incorporation of AI into healthcare has gained global appreciation and AI has been transforming the landscape of patient care and healthcare management. Jacob et al. [20] discuss the technological and organizational factors that influence the adoption of mobile health platforms and emphasize the critical role of ease of use and portability requirements. Jacob et al. [20] also note that apps and tools accessible on smartphones serve as important enablers for both patients and healthcare providers. Moreover, their study focuses on clinicians' perspectives regarding the implementation of telehealth tools, highlighting the importance of obtaining informed consent from patients and ensuring that such applications are user-friendly. In addition, adopting a clinician-centric approach is crucial for successful adoption, as it can significantly impact patient engagement and adherence to treatment plans.

Furthermore, use of wearable technologies, such as smartwatches and medical devices, has been increasing in recent years [21]. Hosseini et al. [22] explore the emergence of virtual care approaches and the ways in which they can leverage technologies to monitor and track health conditions. Likewise, Singh et al. [23] unpack the promises and challenges of using AI in community practice, focusing on the use of new technologies like digital stethoscopes to help enhance data-collection during virtual encounters. Smartwatches also offer similar benefits, including the convenience of testing various health parameters from home, timely detection of minor ailments, and monitoring conditions such as COVID-19 [21, 22]. However, several studies raise important concerns regarding the credibility of these devices, pointing out the lack of transparency in the algorithms used for detecting and analyzing health data [24] as well as the need for greater clarity about whether and when a model performs well on specific patient populations [23]. This uncertainty can lead to potential misinterpretations of health information and misapplications of health technologies, which may affect patient safety and, unsurprisingly, patient trust in these technologies.

It is possible that AI will advance in a way that helps us the identification of solutions to some of the challenges described above, however, it is likely that ethical, regulatory and relevant industry standards will need to step in to help fill the gaps that leave users feeling uncertain about AI safety and efficacy. Algorithmic transparency may also never be realized [25] particularly with commercial AI products given the capitalist incentive structures at play in this healthcare space. This is not to be pessimistic about AI, but to note that clear gaps remain in the ways in which developing, testing, deploying and integrating AI that need to be addressed. It is by focusing on both the advantages and limitations of healthcare informatics technologies that clinicians will be better informed and able to understand the impact of these technologies on healthcare delivery and patient outcomes. Innovations continue to evolve, and it is critical for healthcare providers, policymakers, technology developers and patients to collaborate and ensure that these tools are only being used when there is a confidence that they are safe, effective, and aligned with the needs of patient, clinicians and health system workflows.

4.3 Health Information Exchanges

Health Information Exchanges (HIE) are critical in facilitating the distribution of patient data within the healthcare system and can enhance communication and coordination among healthcare providers [26]. Cochran et al. [27] examine healthcare providers' perspectives on HIE, emphasizing that accessibility of laboratory reports and prescriptions are vital for HIE's acceptance. Accessibility to various features not only streamline clinical workflows but also empower healthcare providers to make more informed decisions based on comprehensive patient data. Most healthcare professionals surveyed by [27] expressed a strong belief that HIE could significantly reduce the need for repetitive clinical tests, thereby improving the efficiency of patient care and reducing overall healthcare costs [27].

However, HIE faces several challenges to its widespread implementation. For instance, Cochran et al. [27] identify cost as a primary barrier and the financial investments required to establish, maintain, and upgrade HIE systems can be substantial, particularly for smaller practices and rural healthcare providers. The integration of HIE into existing clinical workflows also poses difficulties, particularly when it comes to training staff to use these systems effectively. Moreover, Cochran et al. [27] argue that it is crucial to tailor HIE implementations to align with the specific needs and capabilities of end users, including clinicians and administrative staff, to maximize the benefits and to ensure successful adoption.

Rosenkrantz et al. [19] further identify poor design and usability as significant factors contributing to the limited adoption of HIE in clinical practice. Their study underscores that many HIE platforms suffer from cumbersome interfaces and complex functionalities that can frustrate users rather than facilitate their work. The difficulties encountered in interoperability between different institutions are highlighted as a critical limitation, which can lead to delays in diagnosis and treatment. Moreover, interoperability issues can create barriers to timely patient care and erode trust among providers regarding the reliability of the information exchanged [28].

The combined findings from these studies underscore the need for a comprehensive approach to HIE development that prioritizes user-centered design, streamlined workflows, and cost-effectiveness. Enhancing the usability and functionality of HIE systems, while also addressing financial and training barriers, is essential for fostering greater acceptance among healthcare providers. Moreover, this can unlock the full potential of HIE to facilitate integrated, patient-centered care and improve health outcomes across healthcare systems globally.

4.4 Health Informatics in Specific Settings

Health informatics is increasingly critical in managing patient data and facilitating medication reconciliation within primary healthcare settings. According to [29] medication reconciliation is recognized as a crucial safety tool that enables primary care providers to systematically track and manage patient medications. This process is crucially important for minimizing the risk of medication errors, which can occur due to incomplete information or miscommunication during transitions in patient care. Furthermore, primary care providers regard medication reconciliation as a critical component of their practice, as it promotes patient safety and ensures continuity of care, especially for individuals with complex medication regimens [30].

Integration of electronic medical records (EMRs) and health information exchanges in emergency rooms (ER) and operating rooms (OR) can significantly enhance the quality of treatment provided to patients. EMR systems provide prompt access to comprehensive patient data, including past medical histories, allergies, and current medications, which is critical in areas like the ER and OR where quick decision-making is needed. However, Rosenkrantz et al. [19] highlight that perceptions regarding the credibility and reliability of these systems can vary among radiologists and emergency department personnel. Critical factors such as system design, usability, and the flow of information can impact healthcare professionals' trust in these technologies. For instance, if an EMR system is perceived as cumbersome or prone to errors, clinicians may hesitate to rely on it during critical moments, potentially compromising patient care. This underscores the need for continued improvements in the design and functionality of EMRs and HIEs, seeking and incorporating provider feedback, as well as enhanced training for healthcare providers to build confidence in these newer tools.

Additionally, Senteio et al. [31] discuss the significant limitations of electronic health records (EHRs) regarding their ability to capture and retrieve psychosocial information, which is essential for the holistic management of chronic conditions such as hypertension and diabetes mellitus. Psychosocial factors including mental health status, social support systems, and lifestyle factors play a crucial role in effective hypertension and diabetes management. The study points out that the current structure of many EHRs often hinders the efficient retrieval of this vital information, resulting in fragmented care and potentially poor treatment outcomes. To address these challenges, it is crucial to remodel EHR systems to facilitate the collection and integration of psychosocial data. Such improvements could include creating standardized fields for psychosocial assessments, enhancing user interfaces for easier data entry and retrieval, and implementing decision-support tools that consider psychosocial factors in treatment plans.

Furthermore, health informatics systems such as EMRs and HIEs offer substantial benefits for managing patient data and improving care quality, however, there are significant challenges that must be addressed. Prioritizing user-centered design, enhancing the credibility of systems in high-pressure settings such as the ER and OR, and integrating comprehensive psychosocial information are essential steps toward optimizing these technologies. By increasing focus on these areas, healthcare organizations can better support clinicians in delivering safe, effective, and patient-centered care.

Confidence is also crucial for the successful adoption of health informatics tools. This refers to the proficiency and comfort level that healthcare practitioners feel when utilizing these systems. Lack of confidence can lead to underutilization or misuse of informatics tools, ultimately impacting patient care. Training programs that focus on hands-on learning experiences, simulations, and practical applications can help build this confidence. For instance, incorporating real-world scenarios into training can allow practitioners to practice decision-making in a controlled environment, thereby increasing their comfort level with the technology [32]. Accountability is another critical factor, emphasizing the responsibility of users to ensure data integrity and maintain patient safety [32, 33]. Accountability includes adherence to ethical guidelines, data protection regulations, and best practices for information management [33]. Training programs that clearly outline these responsibilities, alongside real-life case studies highlighting the consequences of breaches or errors, can enhance practitioners' understanding of their roles in maintaining high standards of care.

Visosky [32] conducted an extensive evaluation of the current state of health informatics training, identifying those three pivotal factors of interaction, confidence, and accountability, significantly influence the experiences of all actors involved in health informatics. Interaction is the collaborative dynamics between healthcare providers, patients, and technology. Effective interaction is essential for ensuring that health informatics tools are used to their full potential. Achieving effective interaction requires fostering communication among team members, integrating feedback mechanisms, and creating an environment where practitioners feel comfortable sharing their experiences and challenges with these technologies. Engaging training programs that emphasize teamwork and collaborative problem-solving can enhance user interaction, leading to better outcomes [34].

Complementing these findings, Morgan et al. [35] explored primary care practitioners' perceptions regarding the accessibility, visual design, and various characteristics of an interactive health informatics tool, the "Whole Day Matters Tool and User Guide," aimed at facilitating discussions about healthy movement behaviors among practitioners and patients. The study underscores that the usability of such tools directly influences healthcare providers' ability to effectively incorporate them into their practices. Key aspects like intuitive visual design, clear navigation, and easy access to relevant information are critical for enhancing user experience and promoting adherence to patient care protocols.

Furthermore, programs that specifically address challenges within health informatics training can significantly improve healthcare providers' openness to and proficiency in utilizing these tools [36]. For instance, incorporating targeted training sessions that focus on overcoming common barriers such as resistance to change or fear of technology can lead to increased engagement and better utilization of informatics systems [2]. Equally important, the careful planning and implementation of change management strategies at the institutional level could improve user readiness to adopt new technologies [37]. Moreover, programming or training to help promote peer learning and knowledge sharing can also be beneficial, allowing practitioners to learn from each other's experiences and best practices. By addressing critical issues such as interaction, confidence, and accountability in health informatics training, educational programs will better equip healthcare providers with the necessary skills and knowledge to navigate the complexities of modern healthcare technology. This comprehensive approach not only enhances user proficiency but ultimately contributes to improved patient outcomes and more efficient systems of healthcare delivery.

5 Conclusion

This chapter highlights the diverse perspectives of healthcare providers regarding health informatics, emphasizing both the potential benefits and the ongoing challenges associated with its implementation. Even though emergence of health informatics presents numerous advantages, such as improved patient care, enhanced communication, and streamlined workflows, significant barriers to acceptance and concerns over data security persist. Healthcare providers recognize the potential for health informatics to transform clinical practice by facilitating real-time access to and analysis of patient data, enhancing decision-making capabilities, and promoting collaborative care models. Moreover, these technologies can lead to more personalized treatment plans and improved health outcomes, as well as increased efficiency in managing administrative tasks. However, despite these advantages, many practitioners' express hesitation about fully adopting these systems, often due to their apprehensiveness about the security of sensitive patient data and the potential for breaches that could compromise patient confidentiality. Continuous education and skill development are critical for empowering healthcare providers to utilize health informatics tools confidently. Training programs should be tailored to address the specific needs and concerns of practitioners, focusing on practical applications and the ethical implications of data management. By enhancing the comfort level of users with technology, providers are more likely to embrace these tools in their practice. Moreover, solid technological infrastructure is vital for the successful implementation of health informatics systems. This includes not only the necessary hardware and software but also reliable data security measures to protect patient information. Investing in advanced cybersecurity protocols and ensuring that systems are interoperable can alleviate many concerns surrounding data safety and accessibility.

Furthermore, establishing comprehensive and clear institutional policies that govern the use of health informatics is essential for fostering a culture of accountability and trust. Policies should outline evidence-informed practices for data management, confidentiality, and compliance with regulatory standards. Engaging users and other actors affected by these technologies in the development of these policies can ensure that they are practical, protective and reflective of the needs of both healthcare providers and their patients. In conclusion, while the potential benefits of health informatics are significant, addressing the challenges of acceptance, adoption and data security requires a comprehensive strategy. By focusing on training, infrastructure, and institutional policies, healthcare organizations can create an environment that not only supports the adoption of health informatics but also safeguards the integrity and confidentiality of patient data, ultimately leading to improved healthcare delivery and outcomes.

References

1. Bajwa, J., Munir, U., Nori, A., Williams, B.: Artificial intelligence in healthcare: transforming the practice of medicine. Future Healthc. J. **8**(2), e188–e194 (2021). https://doi.org/10.7861/fhj.2021-0095
2. Yogesh, M.J., Karthikeyan, J.: Health informatics: engaging modern healthcare units: a brief overview. Front. Public Health **10**, 854688 (2022). https://doi.org/10.3389/fpubh.2022.854688
3. Li, Y., Albarrak, A.S.: An informatics-driven intelligent system to improve healthcare accessibility for vulnerable populations. J. Biomed. Inform. **134**, 104196–104196 (2022). https://doi.org/10.1016/j.jbi.2022.104196
4. Sheikh, A. (ed.).: Key Advances in Clinical Informatics: Transforming Health Care Through Health Information Technology, 1st edn. Academic Press (2017)
5. Jen, M.Y., Mechanic, O.J., Teoli, D.: Informatics. StatPearls—NCBI Bookshelf (2023). https://www.ncbi.nlm.nih.gov/books/NBK470564/
6. Adam, T., Aliferis, C.: Personalized and Precision Medicine Informatics. Health Informatics Series. Springer, Basel (2020).
7. Chisholm, R.L.: At the interface between medical informatics and personalized medicine: the emerge network experience. Healthc. Inf. Res. **19**(2), 67–68 (2013). https://doi.org/10.4258/hir.2013.19.2.67
8. Singh, K., Kumar, A., Gupta, M.K.: Applications and analytics of bioinformatics, healthcare informatics for modern healthcare system. Revista Română De Informatică Şi Automatică **32**(4), 69–76 (2022). https://doi.org/10.33436/v32i4y202206
9. Brown, S., Hudson, C., Hamid, A., Berman, G., Echefu, G., Lee, K., Lamberg, M., Olson, J.: The pursuit of health equity in digital transformation, health informatics, and the cardiovascular learning healthcare system. Am. Heart J. Plus Cardiol. Res. Practice **17**, 100160 (2022). https://doi.org/10.1016/j.ahjo.2022.100160
10. Himani, K., Surya, J.S.P., Shobhit, A., Pratap, S.A.: Health informatics: an emerging strategy for clinical care. In: Tripathi, G., Shakya, A., Kanga, S., Guite, L.T.S., Singh, S.K. (eds.) Sustainability and Health Informatics. Advances in Geographical and Environmental Sciences. Springer, Singapore (2024). https://doi.org/10.1007/978-981-97-6706-9_13
11. Dash, S., Shakyawar, S.K., Sharma, M., Kaushik, S.: Big data in healthcare: management, analysis and future prospects. J. Big Data **6**(1) (2019). https://doi.org/10.1186/s40537-019-0217-0
12. Eappen, P., Parker Davidson, K., MacLeod, M., Cousins, A., Gunn, V.: Global healthcare informatics—Structural considerations and workforce training challenges and solutions. In: Vajihala, N.R., Eappen. P. (eds.) Healthcare Informatics Innovation Post-COVID-19 Pandemic. Taylor & Francis, Routledge
13. Kirkendall, E.S., Goldenhar, L.M., Simon, J.L., Wheeler, D.S., Spooner, S.A.: Transitioning from a computerized provider order entry and paper documentation system to an electronic health record: Expectations and experiences of hospital staff. Int. J. Med. Inf. **82**(11), 1037–1045 (2013). https://doi.org/10.1016/j.ijmedinf.2013.08.005
14. Gagnon, M., Ngangue, P., Payne-Gagnon, J., Desmartis, M.: M-Health adoption by healthcare professionals: a systematic review. J. Am. Med. Inform. Assoc. **23**(1), 212–220 (2015). https://doi.org/10.1093/jamia/ocv052

15. Mackert, M., Mabry-Flynn, A., Champlin, S., Donovan, E.E., Pounders, K.: Health literacy and health information technology adoption: the potential for a new digital divide. J. Med. Internet Res. **18**(10), e264 (2016). https://doi.org/10.2196/jmir.6349
16. Harish, V., Ackery, A., Grant, K., Jamieson, T., Mehta, S.: Cyberattacks on Canadian health information systems. CMAJ **195**(45), E1548–E1554 (2023). https://doi.org/10.1503/cmaj.230436
17. Alanazi, A.T.: Clinicians' perspectives on healthcare cybersecurity and cyber threats. Cureus **14;15**(10), e47026 (2023). https://doi.org/10.7759/cureus.47026
18. Krousel-Wood, M., McCoy, A.B., Ahia, C., Holt, E.W., Trapani, D.N., Luo, Q., Price-Haywood, E.G., Thomas, E.J., Sittig, D.F., Milani, R.V.: Implementing electronic health records (EHRs): health care provider perceptions before and after transition from a local basic EHR to a commercial comprehensive EHR. J. Am. Med. Inform. Assoc. **25**(6), 618–626 (2017). https://doi.org/10.1093/jamia/ocx094
19. Rosenkrantz, A.B., Smith, S.W., Recht, M.P., Horwitz, L.I.: Perceptions of radiologists and emergency medicine providers regarding the quality, value, and challenges of outside image sharing in the emergency department setting. Am. J. Roentgenol. **214**(4), 843–852 (2020). https://doi.org/10.2214/ajr.19.22096
20. Jacob, C., Sanchez-Vazquez, A., Ivory, C.: Social, organizational, and technological factors impacting clinicians' adoption of mobile health tools: systematic literature review. JMIR Mhealth Uhealth **8**(2), e15935 (2019). https://doi.org/10.2196/15935
21. Eappen, P.: Healthcare Informatics During the COVID-19 Pandemic. In: Advances in Logistics, Operations, and Management Science Book Series, pp. 193–209 (2022). https://doi.org/10.4018/978-1-6684-5279-0.ch010
22. Hosseini, M.M., Hosseini, S.T.M., Qayumi, K., Hosseinzadeh, S., Tabar, S.S.S. Smartwatches in healthcare medicine: assistance and monitoring; a scoping review. BMC Med. Inf. Decis. Mak. **23**(1) (2023a). https://doi.org/10.1186/s12911-023-02350-w
23. Singh, D., Nagaraj, S., Daniel, R., Flood, C., Kulik, D., Flook, R., Goldenberg, A., Brudno, M., Stedman, I.: The promises and challenges of clinical AI in community paediatric medicine. Paediatric Child Health. **28**(4), 212–217 (2023). https://doi.org/10.1093/pch/pxac080
24. Zhang, J., Zhang, Z.: Ethics and governance of trustworthy medical artificial intelligence. BMC Med. Inf. Decis. Mak. **23**(1) (2023). https://doi.org/10.1186/s12911-023-02103-9
25. McCradden, M., Stedman, I.: Explaining decisions without explainability? Artificial intelligence and medicolegal accountability. Future Health. J. **1;11**(3), 100171 (2024). https://doi.org/10.1016/j.fhj.2024.100171
26. Menachemi, N., Rahurkar, S., Harle, C.A., Vest, J.R.: The benefits of health information exchange: an updated systematic review. J. Am. Med. Inform. Assoc. **25**(9), 1259–1265 (2018). https://doi.org/10.1093/jamia/ocy035
27. Cochran, G.L., Lander, L., Morien, M., Lomelin, D.E., Sayles, H., Klepser, D.G.: Health care provider perceptions of a query-based health information exchange: barriers and benefits. J. Innov. Health Inf. **22**(2), 302–308 (2015). https://doi.org/10.14236/jhi.v22i2.135
28. Torab-Miandoab, A., Samad-Soltani, T., Jodati, A., Rezaei-Hachesu, P.: Interoperability of heterogeneous health information systems: a systematic literature review. BMC Med. Inf. Decis. Mak. **23**(1) (2023). https://doi.org/10.1186/s12911-023-02115-5
29. Lesselroth, B.J., Holahan, P.J., Adams, K., Sullivan, Z.Z., Church, V.L., Woods, S., Felder, R., Adams, S., Dorr, D.A.: Primary care provider perceptions and use of a novel medication reconciliation technology. J. Innov. Health In. **19**(2), 105–118 (2011b). https://doi.org/10.14236/jhi.v19i2.802
30. Redmond, P., Grimes, T.C., McDonnell, R., Boland, F., Hughes, C., Fahey, T.: Impact of medication reconciliation for improving transitions of care. Cochrane Library **2018**(8) (2018). https://doi.org/10.1002/14651858.cd010791.pub2
31. Senteio, C., Veinot, T., Adler-Milstein, J., Richardson, C.: Physicians' perceptions of the impact of the EHR on the collection and retrieval of psychosocial information in outpatient diabetes care. Int. J. Med. Informatics **113**, 9–16 (2018). https://doi.org/10.1016/j.ijmedinf.2018.02.003

32. Visosky, P.: How well prepared is the next generation of health professionals to use health care informatics? A qualitative study of stakeholders' views about health informatics curricula in the UAE. The University of Bath's Research Portal (2021). https://researchportal.bath.ac.uk/en/studentTheses/how-well-prepared-is-the-next-generation-of-health-professionals-
33. Mennella, C., Maniscalco, U., De Pietro, G., Esposito, M.: Ethical and regulatory challenges of AI technologies in healthcare: a narrative review. Heliyon **10**(4), e26297 (2024). https://doi.org/10.1016/j.heliyon.2024.e26297
34. Rosen, M.A., DiazGranados, D., Dietz, A.S., Benishek, L.E., Thompson, D., Pronovost, P.J., Weaver, S.J.: Teamwork in healthcare: key discoveries enabling safer, high-quality care. Am. Psychol. **73**(4), 433–450 (2018). https://doi.org/10.1037/amp0000298
35. Morgan, T.L., Pletch, J., Faught, E., Fortier, M.S., Gazendam, M.K., Howse, K., Jain, R., Lane, K.N., Maclaren, K., McFadden, T., Prorok, J.C., Weston, Z.J., Tomasone, J.R.: Developing and testing the usability, acceptability, and future implementation of the Whole Day Matters Tool and User Guide for primary care providers using think-aloud, near-live, and interview procedures. BMC Med. Inf. Decis. Mak. **23**(1) (2023). https://doi.org/10.1186/s12911-023-02147-x
36. Safi, S., Thiessen, T., Schmailzl, K.J.: Acceptance and resistance of new digital technologies in medicine: qualitative study. JMIR Res. Protocols **7**(12), e11072 (2018). https://doi.org/10.2196/11072
37. Harrison, R., Fischer, S., Walpola, R.L., Chauhan, A., Babalola, T., Mears, S., Le-Dao, H.: Where do models for change management, improvement and implementation meet? A systematic review of the applications of change management models in healthcare. J. Health. Leadership **13**, 85–108 (2021). https://doi.org/10.2147/JHL.S289176

Philip Eappen is an Assistant Professor of Healthcare Management at Cape Breton University and serves as the Director of Clinical Services and Transition to Community at the Breton Ability Center (currently on a leave of absence). With over a decade of experience in hospital and healthcare management operations, Dr. Eappen has an extensive academic background. Prior to joining Cape Breton University, he taught healthcare management at the University of Toronto, Southern Alberta Institute of Technology, and Fanshawe College. He has also worked as Director of Health Services and Chief Administrator of Healthcare Operations at the American University of Nigeria and served as a faculty member at other universities and colleges. Dr. Eappen holds a Doctorate in Healthcare Administration from Central Michigan University, a Master of Business Administration in Healthcare Management, a Bachelor of Nursing, and a Post-Graduate Certificate in International Health. He is a board director for various organizations, including the American College of Healthcare Executives, Myeloma Canada, and Aplastic Anemia and Myelodysplasia Canada. Additionally, Dr. Eappen is a long-serving member of the American College of Healthcare Executives and the Canadian College of Health Leaders.

Ian Stedman is an Associate Professor of Canadian Public Law and Governance at York University, with cross-appointments in graduate programs at Osgoode Hall Law School, Science and Technology Studies, and Socio-Legal Studies. He serves on the executive boards of York's Centre for AI & Society and Connected Minds. A former practicing lawyer and public sector advisor, Dr. Stedman specializes in ethics, accountability, and governance. Living with a rare genetic disease, he advocates for the rare disease community and leads research on personalized healthcare technologies and policies. He is Vice Chair of the CIHR Institute of Genetics' Advisory Board and Chair of the Canadian Autoinflammatory Network. Dr. Stedman also contributes as a legal member of SickKids' Research Ethics Board panels and has earned accolades for his work in law and technology, including the IP Osgoode David Vaver Medal for Excellence in IP Law.

Virginia Gunn is an Assistant Professor in the School of Nursing at Cape Breton University, NS, Canada and an Affiliate Researcher at the Institute of Environmental Medicine, Karolinska Institute, Sweden. She completed her Ph.D. and Master studies at the University of Toronto, where she acquired cross-disciplinary academic and research training in health and social sciences. Dr. Gunn is currently involved in several artificial intelligence related international projects involving multidisciplinary teams of researchers from Sweden, Finland, the Netherlands, Spain, Australia, and the US. Specific areas of research include algorithmic management and its impact on workers' health and wellbeing, the application of algorithmic management to health care, and the training of the workforce in healthcare informatics.

The manufacturer's authorised representative in the EU is Springer Nature Customer Service Centre GmbH, Europaplatz 3, 69115 Heidelberg, Germany. If you have any concerns regarding our products, please contact ProductSafety@springernature.com

Printed and bound by CPI Group (UK) Ltd, Croydon, CR0 4YY
30/09/2026
02225558-0009